My
DROID™

SECOND EDITION

Craig James Johnston

800 East 96th Street,
Indianapolis, Indiana 46240 USA

My DROID™, Second Edition

Copyright © 2012 by Pearson Education, Inc.

ISBN-13: 978-0-7897-4938-3
ISBN-10: 0-7897-4938-6

Library of Congress Cataloging-in-Publication Data is on file.

Printed in the United States of America

First Printing: October 2011

Trademarks

All terms mentioned in this book that are known to be trademarks or service marks have been appropriately capitalized. Que Publishing cannot attest to the accuracy of this information. Use of a term in this book should not be regarded as affecting the validity of any trademark or service mark.

DROID is a trademark of Lucasfilm Ltd. and its related companies.

Warning and Disclaimer

Every effort has been made to make this book as complete and as accurate as possible, but no warranty or fitness is implied. The information provided is on an "as is" basis. The author and the publisher shall have neither liability nor responsibility to any person or entity with respect to any loss or damages arising from the information contained in this book.

Bulk Sales

Que Publishing offers excellent discounts on this book when ordered in quantity for bulk purchases or special sales. For more information, please contact

U.S. Corporate and Government Sales

1-800-382-3419

corpsales@pearsontechgroup.com

For sales outside of the U.S., please contact

International Sales

international@pearson.com

EDITOR-IN-CHIEF
Greg Wiegand

ACQUISITIONS EDITOR
Michelle Newcomb

DEVELOPMENT EDITOR
Charlotte Kughen,
The Wordsmithery LLC

MANAGING EDITOR
Kristy Hart

PROJECT EDITOR
Anne Goebel

INDEXER
Lisa Stumpf

PROOFREADER
Kathy Ruiz

TECHNICAL EDITOR
Christian Kenyeres

PUBLISHING COORDINATOR
Cindy Teeters

BOOK DESIGNER
Anne Jones

COMPOSITOR
Tricia Bronkella

Contents at a Glance

Table of Contents

About the Author

Craig James Johnston has been involved with technology since his high school days at Glenwood High in Durban, South Africa, when his school was given some Apple II Europluses. From that moment technology captivated him, and he has owned, supported, evangelized, and written about it.

Craig has been involved in designing and supporting large-scale enterprise networks with integrated email and directory services since 1989. He has held many different IT-related positions in his career ranging from sales support engineer to mobile architect for a 40,000-smartphone infrastructure at a large bank.

In addition to designing and supporting mobile computing environments, Craig writes about it for BYTE Magazine at BYTE.com and CrackBerry.com, a leading BlackBerry blog. Craig also cohosts the CrackBerry.com, BYTE Wireless Radio, and BYTE Apple Radio podcasts. You can see Craig's previously published work in his books *Professional BlackBerry*, *My BlackBerry Curve*, *My Palm Pre*, *My Nexus One*, *My Motorola Atrix 4G*, and *My BlackBerry PlayBook*.

Craig also enjoys high-horsepower, high-speed vehicles and tries very hard to keep to the speed limits while driving them.

Originally from Durban, South Africa, Craig has lived in the United Kingdom, the San Francisco Bay Area, and New Jersey where he now lives with his wife, Karen, and a couple of cats.

Craig would love to hear from you. Feel free to contact Craig about your experiences with *My DROID* at http://www.CraigsBooks.info.

All comments, suggestions, and feedback are welcome, including positive and negative.

Dedication

"Bullying consists of the least competent most aggressive employee projecting their incompetence on to the least aggressive most competent employee and winning."
—Tim Field

Acknowledgments

I would like to express my deepest gratitude to the following people on the *My Droid* team who all worked extremely hard on this book.

Michelle Newcomb, my acquisitions editor who worked with me to give this project an edge, as well as technical editor Christian Keyneres, development editor Charlotte Kughen, project editor Anne Goebel, indexer Lisa Stumpf, compositor Tricia Bronkella, and proofreader Kathy Ruiz.

A special thanks to Kerry Harrington from Weber Shandwick for providing five DROIDs on loan so I could get this second edition in on time.

We Want to Hear from You!

As the reader of this book, *you* are our most important critic and commentator. We value your opinion and want to know what we're doing right, what we could do better, what areas you'd like to see us publish in, and any other words of wisdom you're willing to pass our way.

As an associate publisher for Que Publishing, I welcome your comments. You can email or write me directly to let me know what you did or didn't like about this book—as well as what we can do to make our books better.

Please note that I cannot help you with technical problems related to the topic of this book. We do have a User Services group, however, where I will forward specific technical questions related to the book.

When you write, please be sure to include this book's title and author as well as your name, email address, and phone number. I will carefully review your comments and share them with the author and editors who worked on the book.

Email: feedback@quepublishing.com

Mail: Greg Wiegand
Editor-in-Chief
Que Publishing
800 East 96th Street
Indianapolis, IN 46240 USA

Reader Services

Visit our website and register this book at quepublishing.com/register for convenient access to any updates, downloads, or errata that might be available for this book.

Prologue

In this chapter, you learn about the external features of the DROID and the basics of getting started with your DROID. Topics include the following:

- → Your DROID's external features
- → Fundamentals of Android
- → Installing synchronization software

Getting to Know Your DROID

Let's start by getting to know more about your DROID by examining the external features, device features, and how the Android operating system works.

What Is DROID?

DROID is a clever marketing name given to some smartphones running Android on the Verizon CDMA cellular network in the United States. Android smartphones that are given the DROID name can come from different manufacturers. Many times these same phones are available in GSM form outside the U.S. This edition of *My DROID* covers the DROID 3, DROID X2, DROID Incredible 2, DROID Pro, and the DROID CHARGE. Now that you know what DROID is, let's take a look at the external features of your DROID.

Your DROID's External Features

The external features of the DROID vary slightly from one model to another. For example, some DROIDs have physical keyboards and a camera button and others do not.

DROID 3 and Motorola Milestone 3

The DROID 3 is also called the Motorola Milestone 3 outside of the U.S. All features you see here are the same on both phones.

Light sensor

Indicator light

Earpiece

Front camera

Proximity sensor

Volume up/ down buttons

Touchscreen

Menu button

Search button

Home button   Back button

- **Proximity sensor**   Detects when you place your DROID Pro against your ear to talk, which causes it to turn off the screen to prevent any buttons from being pushed inadvertently.

- **Light sensor**   Adjusts the brightness of your DROID Pro's screen based on the brightness of the ambient light.

- **Earpiece**

- **Indicator light**   Indicates new events (such as missed calls or new email).

- **Front camera**   1.3 Megapixel front-facing camera that can be used for video chat.

- **Touchscreen**   The DROID 3 has a 4.0" 960×540 pixel LCD (Liquid Crystal Display) Thin-film Transistor (TFT) screen with capacitive touch and a white back light.

- **Volume up/down buttons**   Control the audio volume on calls and while you're playing audio and video.

- **Back button**   Touch to go back one screen when using an application or menu. This soft button doesn't actually press in. When you touch it, your DROID 3 vibrates briefly to let you know it has detected the touch.

- **Menu button**   Displays a menu of choices. The menu differs based on what screen you are looking at and what application you are using.

- **Home button**   Touch to go to the Home screen. The application that you are using continues to run in the background.

- **Search button**   Touch to type or speak a search term. Your DROID 3 searches your phone and the Internet for content that matches the search term.

8 Megapixel camera with autofocus

LED camera flash

Noise cancelling microphone

Speaker

- **8 Megapixel camera with autofocus**   Takes clear pictures close-up or far away.

- **LED (Light Emitting Diode) camera flash**   Helps to illuminate the surroundings when taking pictures in low light.

- **Speaker**   Audio is produced when speakerphone is in use. Place your DROID 3 on a hard surface for the best audio reflection.

- **Noise cancelling microphone**   Used in conjunction with the regular microphone on phone calls to reduce background noise.

3.5 mm headphone jack

Power button

Navigation buttons

Full physical keyboard

- **3.5 mm headphone jack**   Plug in your DROID or third-party headsets to enjoy music and talk on the phone.

- **Power button**   Press once to wake up your DROID 3. Press and hold for one second to reveal a menu of choices. The choices enable you to put your DROID into silent mode, airplane mode, or power it off completely.

- **Full physical keyboard**   Slide the keyboard out from under the screen anytime you need to type. The screen automatically switches to landscape mode.

- **Navigation buttons**   While typing, use the navigation buttons to move around the screen.

- **Micro-USB port**   Use the supplied Micro-USB cable to charge your DROID 3 or connect it to a computer to synchronize multimedia and other content.

- **Micro-HDMI port**   HDMI (High Definition Multimedia Interface) has become the standard for connecting high-definition equipment such as plasma TVs and Blu-Ray players. This HDMI port enables you to play movies on your HDTV.

DROID Pro

Power button   Earpiece

Proximity sensor

Light sensors

Touchscreen

Right-side convenience button

Home button

Back button

Menu button

Search button

Full keyboard

- **Proximity sensor**   Detects when you place your DROID Pro against your ear to talk, which causes it to turn off the screen to prevent any buttons from being pushed inadvertently.

- **Light sensors**   Adjust the brightness of your DROID Pro's screen based on the brightness of the ambient light.

- **Earpiece**

- **Full keyboard**   Full physical keyboard for fast typing.

- **Touchscreen**   The DROID Pro has a 3.1" 320×480 pixel LCD (Liquid Crystal Display) Thin-film Transistor (TFT) screen with capacitive touch and a white back light. Because the screen uses capacitive touch, you do not need to press hard.

- **Right-side convenience button**   Press to open an application you use frequently, such as your calendar. You can customize which application this button opens.

- **Back button**   Touch to go back one screen when using an application or menu. This soft button doesn't actually press in. When you touch it, your DROID Pro vibrates briefly to let you know it has detected the touch.

- **Menu button**   Displays a menu of choices. The menu differs based on what screen you are looking at and what application you are using.

- **Home button**   Touch to go to the Home screen. The application that you are using continues to run in the background.

- **Search button**   Touch to type or speak a search term. Your DROID Pro searches your phone and the Internet for content that matches the search term.

- **Power button**   Press once to wake up your DROID Pro. Press and hold for one second to reveal a menu of choices. The choices enable you to put your DROID into silent mode, airplane mode, or power it off completely.

3.5 mm headphone jack

5 Megapixel camera with autofocus

Dual LED camera flash

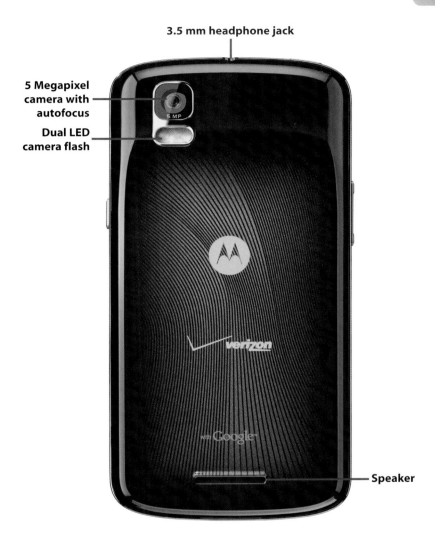

Speaker

- **5 Megapixel camera with autofocus**   Takes clear pictures close-up or far away.

- **Dual LED (Light Emitting Diode) camera flash**   Helps to illuminate the surroundings when taking pictures in low light.

- **Speaker**   Audio is produced when speakerphone is in use. Place your DROID Pro on a hard surface for the best audio reflection.

- **3.5 mm headphone jack**   Plug in your DROID or third-party headsets to enjoy music and talk on the phone.

Volume up/down buttons

Micro-USB port

- **Volume up/down buttons** Controls all audio volume including while you're playing music, watching a video, or talking on the phone.

- **Micro-USB port** Use the supplied Micro-USB cable to charge your DROID Pro or connect it to a computer to synchronize multimedia and other content.

DROID CHARGE

The DROID CHARGE is from Samsung and is the first 4G DROID smartphone offered on Verizon.

Earpiece
Light sensors
Proximity sensor
Volume up/down button
Touchscreen
Menu button
Search button
Home button
Back button

- **Proximity sensor**   Detects when you place your DROID CHARGE against your ear to talk, which causes it to turn off the screen to prevent any buttons from being pushed inadvertently.

- **Light sensors**   Adjust the brightness of your DROID CHARGE's screen based on the brightness of the ambient light.

- **Earpiece**

- **Volume up/down button**   Controls audio while playing music, watching a video, or talking on the phone.

- **Touchscreen**   The DROID CHARGE has a 4.3" 480×800 pixel Super AMOLED (Active Matrix Organic Light Emitting Diode) screen with capacitive touch, which does not require backlight. Consequently, it saves battery life and produces vivid colors. Because the screen uses capacitive touch, you do not need to press hard.

- **Back button**   Touch to go back one screen when using an application or menu. This soft button doesn't actually press in. When you touch it, your DROID CHARGE vibrates briefly to let you know it has detected the touch.

- **Menu button**   Displays a menu of choices. The menu differs based on what screen you are looking at and what application you are using.

- **Home button**   Touch to go to the Home screen. The application that you are using continues to run in the background.

- **Search button**   Touch to type or speak a search term. Your DROID CHARGE searches your phone and the Internet for content that matches the search term.

8 Megapixel camera with autofocus

LED camera flash

Speaker

- **8 Megapixel camera with autofocus**   Takes clear pictures close-up or far away.

- **LED (Light Emitting Diode) camera flash**   Helps to illuminate the surroundings when taking pictures in low light.

- **Speaker**   Audio is produced when speakerphone is in use. Place your DROID CHARGE on a hard surface for the best audio reflection.

Power button

Micro-HDMI port

- **Power button**   Press once to wake up your DROID CHARGE. Press and hold for one second to reveal a menu of choices. The choices enable you to put your DROID CHARGE into silent mode, airplane mode, or power it off completely.

- **Micro-HDMI port**   HDMI (High Definition Multimedia Interface) has become the standard for connecting high-definition equipment such as plasma TVs and Blu-Ray players. This HDMI port enables you to play movies on your HDTV.

Motorola DROID X2

The Motorola DROID X2 does not seem to have a non-U.S. counterpart, so this phone will likely be a Verizon U.S. exclusive.

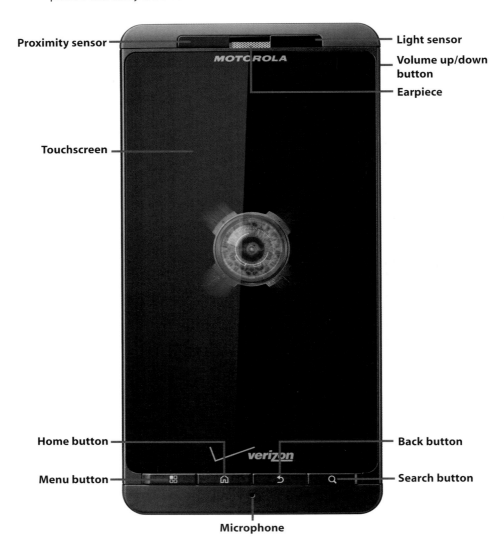

Proximity sensor

Light sensor

Volume up/down button

Earpiece

Touchscreen

Home button

Back button

Menu button

Search button

Microphone

- **Proximity sensor**   Detects when you place your DROID X2 against your ear to talk, which causes it to turn off the screen to prevent any buttons from being pushed inadvertently.

- **Light sensor**   Adjusts the brightness of your DROID X2's screen based on the brightness of the ambient light.

- **Earpiece**

- **Volume up/down button**   Controls audio while playing music, watching a video, or talking on the phone.

- **Touchscreen**   The DROID X2 has a 4.3" 960×540 pixel Thin-film Transistor (TFT) Liquid Crystal Display (LCD) screen with capacitive touch.

- **Microphone**   Picks up your voice when you are on the phone.

- **Back button**   Touch to go back one screen when using an application or menu. This soft button doesn't actually press in. When you touch it, your DROID X2 vibrates briefly to let you know it has detected the touch.

- **Menu button**   Displays a menu of choices. The menu differs based on what screen you are looking at and what application you are using.

- **Home button**   Touch to go to the Home screen. The application that you are using continues to run in the background.

- **Search button**   Touch to type or speak a search term. Your DROID X2 searches your phone and the Internet for content that matches the search term.

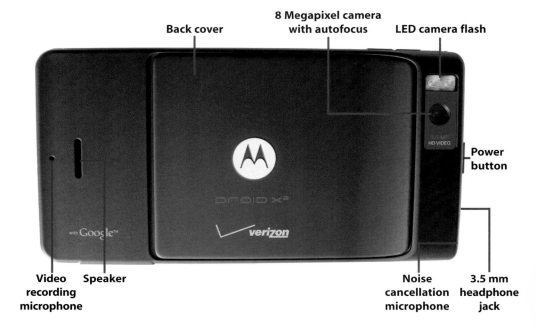

- **8 Megapixel camera with autofocus**   Takes clear pictures close-up or far away.

- **LED (Light Emitting Diode) camera flash**   Helps to illuminate the surroundings when taking pictures in low light.

- **Speaker**   Audio is produced when speakerphone is in use. Place your DROID X2 on a hard surface for the best audio reflection.

- **Back cover**   Press firmly on the back cover to slide it downward when you need to swap the battery or Micro-SD card.

- **Video recording microphone**   When recording videos, this microphone is used.

- **Noise cancellation microphone**   While on a call, this microphone is used to determine background noise and eliminate it.

- **Power button**   Press once to wake up your DROID X2. Press and hold for one second to reveal a menu of choices. The choices enable you to put your DROID X2 into silent mode, airplane mode, or power it off completely.

- **3.5 mm headphone jack**   Plug in your DROID or third-party headsets to enjoy music and talk on the phone.

Micro-USB port —

Micro-HDMI port —

- **Micro-USB port** Use the supplied Micro-USB cable to charge your DROID X2 or connect it to a computer to synchronize multimedia and other content.

- **Micro-HDMI port** HDMI (High Definition Multimedia Interface) has become the standard for connecting high-definition equipment such as plasma TVs and Blu-Ray players. This HDMI port enables you to play movies on your HDTV.

DROID Incredible 2 and HTC Incredible S

The HTC DROID Incredible 2 is also known as the HTC Incredible S outside the U.S. All features you see below are the same for both devices.

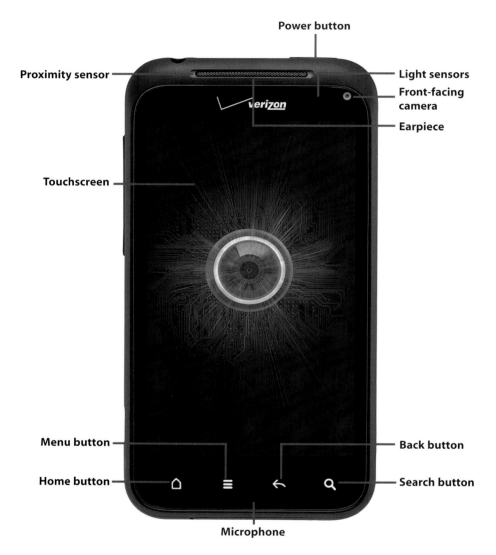

Power button

Proximity sensor

Light sensors

Front-facing camera

Earpiece

Touchscreen

Menu button

Back button

Home button

Search button

Microphone

- **Proximity sensor**   Detects when you place your DROID Incredible 2 against your ear to talk, which causes it to turn off the screen to prevent any buttons from being pushed inadvertently.

- **Light sensors**   Adjust the brightness of your DROID Incredible 2's screen based on the brightness of the ambient light.

- **Front-facing camera**   A 1.3 megapixel camera for taking self portraits or using video chat.

- **Earpiece**

- **Power button**   Press once to wake up your DROID Incredible 2. Press and hold for one second to reveal a menu of choices. The choices enable you to put your DROID into silent mode, airplane mode, or power it off completely.

- **Touchscreen**   The DROID Incredible 2 has a 4" 480×800 pixel AMOLED (Active Matrix Organic Light Emitting Diode) screen with capacitive touch, which does not require backlight and therefore saves on battery life and produces vivid colors.

- **Microphone**   Picks up your voice when you are on the phone.

- **Back button**   Touch to go back one screen when using an application or menu. This soft button doesn't actually press in. When you touch it, your DROID Incredible 2 vibrates briefly to let you know it has detected the touch.

- **Menu button**   Displays a menu of choices. The menu differs based on what screen you are looking at and what application you are using.

- **Home button**   Touch to go to the Home screen. The application that you are using continues to run in the background.

- **Search button**   Touch to type or speak a search term. Your DROID Incredible 2 searches your phone and the Internet for content that matches the search term.

Magic Buttons?

The HTC Incredible S and DROID Incredible 2 have a unique feature. The Menu, Home, Search, and Back buttons change their orientation to match the orientation of the phone.

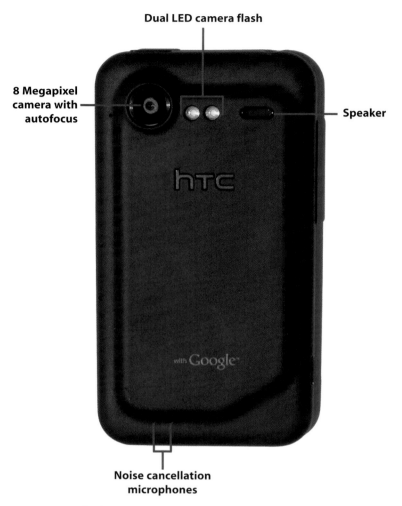

Dual LED camera flash

8 Megapixel camera with autofocus

Speaker

Noise cancellation microphones

- **8 Megapixel camera with autofocus**   Takes clear pictures close-up or far away.

- **Dual LED (Light Emitting Diode) camera flash**   Helps to illuminate the surroundings when taking pictures in low light.

- **Speaker**   Audio is produced when speakerphone is in use. Keep your DROID Incredible 2 on a hard surface for the best audio reflection.

- **Noise cancellation microphones**   While on a call, these microphones are used to determine background noise and eliminate it.

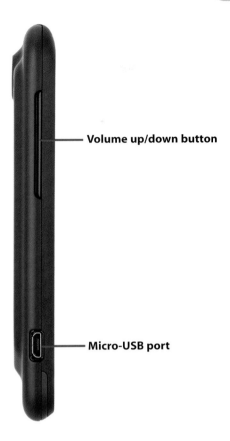

Volume up/down button

Micro-USB port

- **Micro-USB port**   Use the supplied Micro-USB cable to charge your DROID Incredible 2 or connect it to a computer to synchronize multimedia and other content.

- **Volume up/down button**   Controls audio while playing music, watching a video, or talking on the phone.

Fundamentals of Android

Your DROID is run by an operating system called Android. Android was created by Google to run on any smartphone, and there are quite a few that run on Android today. Android works pretty much the same on all smartphones with small differences based on the version of Android installed and the interface tweaks that each manufacturer may have made. Let's go over how to use Android.

The Unlock Screen

If you haven't used your DROID for a while, the screen goes blank to conserve battery power. To unlock your DROID, do the following:

1. Press the power button.

2. Slide the padlock button to the right. This unlocks your DROID.

Silence Your DROID

On the unlock screen, there is a second button that enables you to silence your DROID without having to first unlock it. Simply slide the speaker button to the left to toggle between silent mode and audio alert mode. This does not work on the DROID CHARGE and Incredible 2.

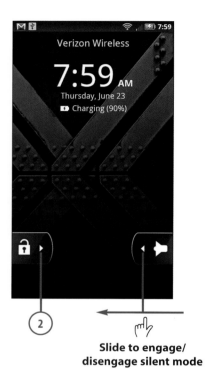

**Slide to engage/
disengage silent mode**

Silence Your DROID 3

Your DROID 3 is a little different. It has a small drag area on the right of the unlock screen that enables you to enable and disable sound. When sound is disabled, your DROID still vibrates when there are alerts.

DROID Incredible 2 Is a Little Different

On most HTC devices running Android, to unlock your DROID, press the power button and drag down on the bar that appears on screen.

Samsung DROID CHARGE Does It Its Own Way

To unlock your DROID CHARGE, slide the puzzle piece into the open puzzle piece slot.

Slide to enable sound on the DROID 3

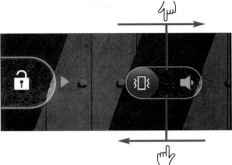

Slide to disable sound on the DROID 3

Slide down to unlock the DROID Incredible 2

Slide the puzzle piece here to unlock the DROID CHARGE

The Home Screen

After you unlock your DROID, you are presented with the Home screen. The Home screen contains application icons, a Launcher icon, status bar, and widgets.

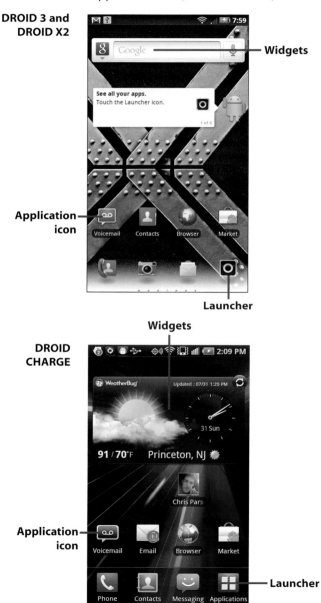

DROID 3 and DROID X2

Widgets

Application icon

Launcher

Widgets

DROID CHARGE

Application icon

Launcher

- **Widgets**   Applications that run right on the Home screen. They are specially designed to provide functionality and real-time information to the Home screen. An example of a widget is one that shows the current weather or provides a search capability.

- **Application icon**   These are application icons that have been dragged to the Home screen from the Launcher. When you touch one, it launches the associated application.

- **Launcher**   Touch to show application icons for all applications that you have installed on your DROID.

Other Home Screen Styles

The DROID Pro and DROID Incredible 2 deviate slightly from the standard Android look and feel of the Home screen as they have their Launcher icons in different places.

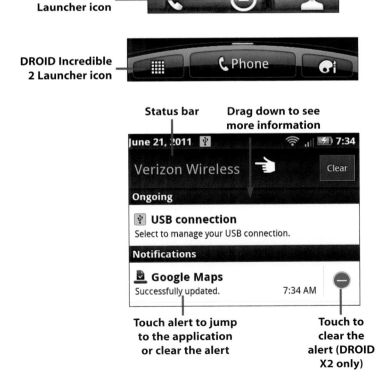

DROID Pro Launcher icon

DROID Incredible 2 Launcher icon

Status bar   Drag down to see more information

Touch alert to jump to the application or clear the alert

Touch to clear the alert (DROID X2 only)

- **Status bar** Shows the time, signal strength, battery level, and which type of network you are connected to, Wi-Fi or cellular. Also shows any notification icons such as new mail.

Drag the Status Bar

You can drag the status bar down to see more notifications or more details on the notifications. When the status bar has been pulled down, touch a notification to be taken to the appropriate application. On the DROID X2 you can touch the red icon to the right of each alert to clear that alert without jumping to the application that created the alert first.

Extra Status Bar Functionality on Some DROIDs

If you have a DROID CHARGE, when you pull down the status bar you see Quick Settings icons at the top that enable you to do things like turn Wi-Fi or Bluetooth on and off. If you have a DROID Incredible 2, you see a list of running applications. Touch one to switch to it. At the bottom you also see a Quick Settings tab that enables you to do things such as turn your Wi-Fi or Bluetooth radios on and off.

DROID CHARGE Quick Settings

DROID Incredible 2
Running Apps

DROID Incredible 2
Quick Settings

Your DROID's Home screen is actually composed of multiple parts or multiple screens. The number of Home screens depends on the device itself. For example, the DROID X2 has seven screens but most Android phones have only five. To see the other parts or screens, touch your finger on the right part of the screen and swipe to the left. You then see another part of the Home screen.

Many DROIDs do not come with any widgets or icons on these other parts of the Home screen, leaving them blank for you to populate.

JUMP BETWEEN HOME SCREENS

You can quickly jump between Home screens by bringing up the thumbnail view of all of the screens. To do this, use the pinch gesture on the Home screen. You see thumbnails of all of the screens. Touch the thumbnail of the screen you want to switch to. On the DROID X2 and DROID 3, you have to swipe up from the bottom of the screen to show the thumbnails.

Using Your DROID's Touchscreen

Interacting with your DROID is done mostly by touching the screen, what's known as making gestures on the screen. You can touch, swipe, pinch, double-tap, and type.

- **Touch**   To start an application, touch its icon. Touch a menu item to select it. Touch the letters of the onscreen keyboard to type.

- **Touch and hold**   Touch and hold to interact with an object. For example, if you touch and hold a blank area of the Home screen, a menu pops up. If you touch and hold an icon, you can reposition it with your finger.

- **Drag**   Dragging always starts with a touch and hold. For example, if you touch the status bar, you can drag it down to read all of the status messages.

- **Swipe or slide**   Swipe or slide the screen to scroll quickly. To swipe or slide, move your finger across the screen quickly. Be careful not to touch and hold before you swipe or you will reposition something.

- **Double-tap**   Double-tapping is like double-clicking a mouse on a desktop computer. Tap the screen twice in quick succession. You can double-tap a web page to zoom in to part of that page. Double-tap also works in the camera application. Double-tap to zoom the camera.

- **Pinch**   To zoom in and out of images and pages, place your thumb and forefinger on the screen. Pinch them together to zoom out or spread them apart to zoom in (unpinching). Applications like Browser, Gallery, and Maps currently support pinching.

- **Rotate the screen**   If you rotate your DROID from an upright position to being on its left side, the screen switches from portrait view to landscape view. Most applications honor the screen orientation. The Home screen does not.

Using Your DROID's Keyboard

With the exception of the Motorola DROID Pro, your DROID has a virtual or onscreen keyboard for those times when you need to enter text. You may be a little wary of a keyboard that has no physical keys, but you will be pleasantly surprised at how well it works. Let's go through the main points of the keyboard.

Some applications automatically show the keyboard when you need to enter text. If the keyboard does not appear, touch the area where you want to type and the keyboard slides up ready for use.

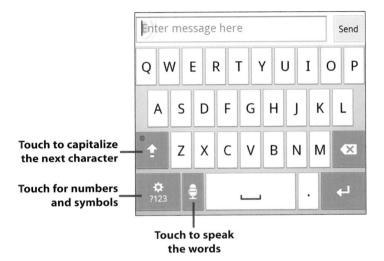

Touch to capitalize the next character

Touch for numbers and symbols

Touch to speak the words

The DROID Pro has a physical keyboard. Use the physical keyboard to type while in portrait mode. If you rotate your DROID Pro on its side, and the application you are using supports screen rotation, you can use the onscreen keyboard.

Hold ALT to type characters in blue

Click to speak the words

Hold ALT and SYM to type symbols

The DROID 3 also has a physical keyboard but it slides out from under the screen. Use the physical keyboard to type while in landscape mode.

Hold ALT to type characters in orange

Navigate the screen and select objects

Click to speak the words

Hold ALT and SYM to type symbols

Using the virtual keyboard as you type, your DROID makes word suggestions. Think of this as similar to the spell checker you would see in a word processor. Your DROID uses a dictionary of words to guess what you are typing. If the word you were going to type is highlighted, touch space or period to select it. If you can see the word in the list but it is not highlighted, touch the word to select it. This feature works when using the on-screen keyboard or the physical keyboard on the DROID Pro.

List of suggested words

To make the next letter you type a capital letter, touch the Shift key. To make all letters capitals (or CAPS), touch and hold or double-tap the Shift key to engage CAPS Lock. Touch Shift again to disengage CAPS Lock.

To backspace or delete what you have typed, touch the DEL key.

To type numbers or symbols on the virtual keyboard, touch the Symbols key. On the physical keyboard on the DROID Pro, hold the ALT key while pressing space to see Symbols.

Add Your Word

If you want to save a suggested word to your dictionary, touch and hold the word. You see a pop-up that shows the word and Saved next to it.

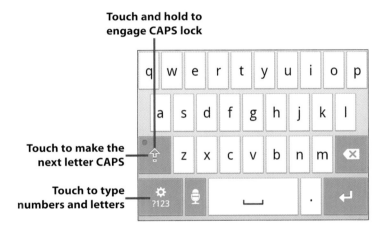

Touch and hold to engage CAPS lock

Touch to make the next letter CAPS

Touch to type numbers and letters

When on the Number and Symbols screen, touch the ALT key to see extra symbols. Touch the ABC key to return to the regular keyboard. From the extra symbols keyboard, touch ALT to return to the numeric keyboard or ABC to see the regular keyboard.

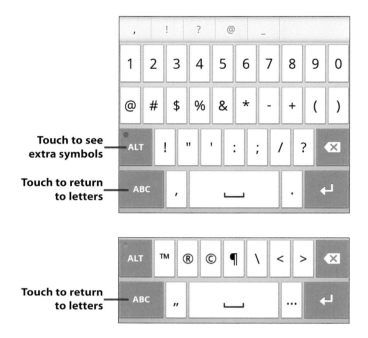

Touch to see extra symbols

Touch to return to letters

Touch to return to letters

Quick Access to Symbols

If you want to type commonly used symbols, touch and hold the period key. A small window opens with those common symbols. Press the X to return to the regular keys.

To enter an accented character, touch and hold any vowel or C, N, or S keys. A small window opens enabling you to select an accented or alternative character. Touch the X to close the window.

Touch a character to type it

Touch and hold for accented characters

Touch to close the accented characters

To reveal other alternative characters, touch and hold any other letter, number, or symbol.

Want a Larger Keyboard?

Turn your DROID sideways to switch to a landscape keyboard. The landscape keyboard has larger keys and is easier to type on.

Speak Instead of Typing

Your DROID can turn your voice into text. It uses Google's speech recognition service, which means that you must have a connection to the cellular network or a Wi-Fi network to use it.

1. Touch or press the microphone key. Alternatively, you can swipe your finger across the onscreen keyboard from left to right.

2. The microphone pops up. Wait for the message "Speak now" and start speaking what you want to be typed. You can speak the punctuation by saying "comma," "question mark," "exclamation mark," or "exclamation point."

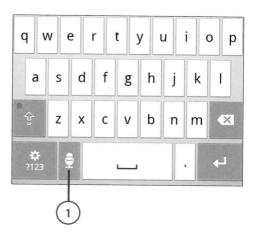

**Start speaking
when you see this**

Touch to cancel

3. After you have spoken what you want to type, you briefly see "working" and then your text is placed in the text field. The text remains underlined so you can touch the DEL key to delete it. If you are satisfied with the text, continue typing on the keyboard or touch the Back button to close the keyboard.

Editing Text—DROID 3, DROID X2, DROID Pro, and DROID CHARGE

After you enter text, you can edit it by cutting, copying, or pasting the text. Here is how to select and copy text.

1. While you are typing, touch and hold the text area.

2. Touch Select Text when the window pops up.

3. Drag the end markers (the little triangles) left and right to highlight the text you want to select.

4. Touch and hold in the text area again after you have highlighted the text.

5. Touch Copy.

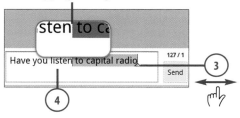

Text around the marker you are dragging is magnified

Editing Text—DROID Incredible 2

After you enter text, you can edit it by cutting, copying, or pasting the text. Here is how to select and copy text.

1. While you are typing, touch and hold the text area.

2. Move your finger along the text to the beginning of a word you want to select and release it.

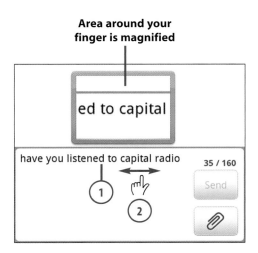

Area around your finger is magnified

3. Touch Select text.

4. Move the green end markers left and right to select more or less text.

5. Touch Copy.

What Can You Do with the Copied Text?

After you have copied some text, you can paste it into any application. To do this, touch and hold the screen where you want to paste the text. When the menu pops up, touch Paste.

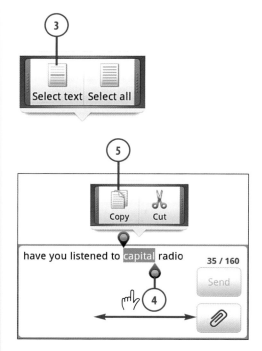

Starting Applications and Application Groups

Applications for your DROID are covered in Chapter 10, "Working with Android Applications," but to get you started, here is how to start an application and switch between applications. Each application has an associated icon. Those icons can be on the Home screen or in the Launcher. The Launcher displays every application you have installed.

1. Touch an application icon to launch that application.

2. Touch the shopping bag icon to launch the Android Market. Read more about Android Market in Chapter 10.

3. Touch All Apps to see the and create application groups.

4. Touch All Apps to see all applications.

5. Touch Recent to see only recently used applications.

6. Touch Downloaded to see only applications that you have downloaded. (Apps shown here do not include the apps that were installed on your DROID when you purchased it.)

7. Touch New Group to add a new application group.

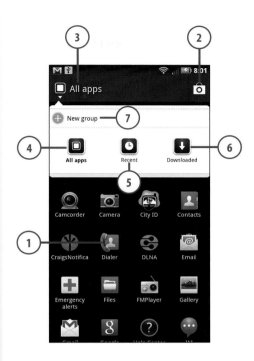

DROID Incredible 2 Groups

After you touch the Launcher icon on your Incredible 2, you'll see that it looks a little different from the other DROIDs. You switch the groups at the bottom of the screen, and when you touch the Menu button, you are able to toggle between an icon and list view of your applications.

Application groups

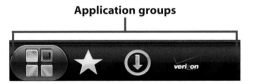

Creating An Application Group

You can create new applications groups to organize your applications and make them easier to find.

1. Touch to add a new group.

2. Type in the group name.

3. Touch to select a group icon.

4. Touch a group icon.

5. Touch to save the group.

6. Touch to add applications to your new group.

Switch Between Applications

You can quickly switch between recently used applications by touching and holding the Home button. A small window pops up revealing the most recently used applications. Touch an icon to switch to that application.

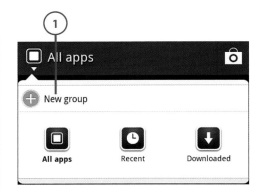

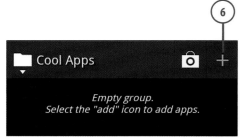

Menus

Your DROID has two types of menus: Options menus and Context menus. Let's go over what each one does.

Most applications have Options menus. These enable you to make changes or take actions within that application. Sometimes the Options menu has a More item that enables you to see more options.

**Touch the Menus button to
reveal the Options Menu**

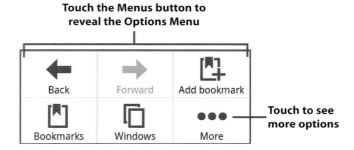

Back   Forward   Add bookmark

**Touch to see
more options**

Bookmarks   Windows   More

A Context menu applies to an item on the screen. If you touch and hold something on the screen (in this example, a link on a web page), a Context menu appears. The items on the Context menu differ based on the type of object you touched.

BYTE Magazine

http://www.google.com/m/
url?client=ms-android-verizon...

Open

Open in new window

Bookmark link

Select text

Copy link URL

Share link

Save link

**Context-appropriate
commands appear when
you touch and hold
something on screen**

Installing Synchronization Software

Your DROID is designed to work in a disconnected fashion without the need to connect it to your desktop computer. However you might still want to synchronize some content from your computer to your DROID. One of the most common uses for this software is to synchronize music and photos. In this book I use an application called doubleTwist. Other applications also provide synchronization for your DROID, such as Missing Sync from Mark/Space, but I've used doubleTwist because it is free. Before we begin, download the version you need (Windows or Mac OSX) from http://www.doubletwist.com.

Installing doubleTwist on Windows

1. Double-click the doubleTwist install file. On the first screen, click Install.

2. After the install is complete, make sure that the Launch doubleTwist box is checked.

3. Click Finish.

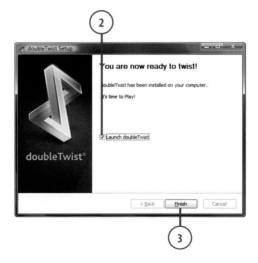

Installing doubleTwist on Mac OSX

1. Double-click on the doubleTwist disc on y our Mac desktop.

2. Drag the doubleTwist icon to the Applications folder icon.

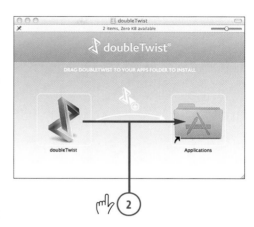

Creating a doubleTwist Account

The first time you run doubleTwist, it asks you to log in using your doubleTwist account. You probably don't have one, so let's go through how to create a free doubleTwist account. The steps and screens are the same for the Windows and Mac versions of doubleTwist.

1. Touch Create Account.

2. Type your name, choose a password, and type your email address.

3. Agree to the doubleTwist EULA.

4. Click Sign Up.

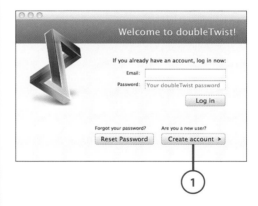

5. The next screen tells you to check your mailbox for an activation email. Switch to your email application and follow the instructions in the email. After your account is activated, switch back to this screen and click Continue to start using doubleTwist.

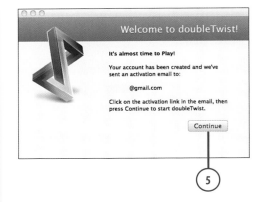

In this chapter, you learn about your DROID's most important application, Contacts. You learn how to add contacts, synchronize contacts, join duplicate contacts together, and even how to add a contact to your Home screen. Topics include the following:

→ Importing contacts
→ Adding contacts
→ Synchronizing contacts
→ Creating favorite contacts

1

Contacts

On any smartphone, including your DROID, the Contacts (People on HTC DROIDs) application is the most important. It is the central hub for many activities such as calling and sending texts (SMS), multimedia files (MMS), or email. You can also synchronize your contacts from many online sites such as Facebook and Gmail so as your friends change their Facebook profile pictures, their picture on your DROID changes, too. Let's open Contacts and look around.

Adding Accounts

The Contacts application runs on the DROID Pro and CHARGE only.

Where Is Contacts on the Other DROIDs?

The DROID X2 and DROID 3 have an icon for Contacts but it really launches another application called Dialer (which is covered later in this chapter). The DROID Incredible 2 has an application called People, which is covered in a different task in this chapter.

Before we look around the Contacts application, let's add some accounts to synchronize contacts from. Due to the differences in the Contacts and People applications used by the different DROIDs, it is easier to add accounts through the Android Settings screens.

Adding a Facebook Account

If you have a Facebook account, you can synchronize your Facebook friends' contact information to your DROID. If someone updates her Facebook profile picture, her contact picture updates on your DROID.

1. From the Home screen, press the Menu button and touch Settings.

2. Touch Accounts. On the DROID Incredible 2, this menu item is labeled Accounts & Sync.

3. Touch Add Account.

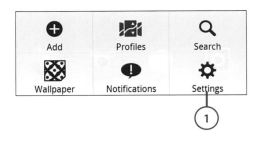

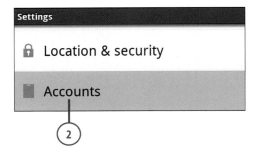

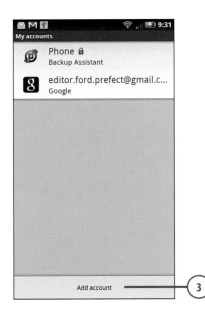

4. Touch Facebook. On the DROID Incredible 2 it is called Facebook for HTC Sense.

5. Enter the email address you used when you signed up for Facebook.

6. Enter your Facebook password.

7. Touch Next. Touch Sign in on the DROID Incredible 2.

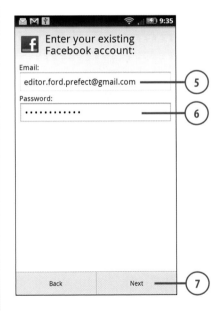

8. Touch Done.

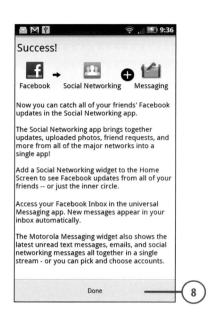

Adding a Work Email Account

Your DROID can synchronize your contacts from your work email account as long as your company uses Microsoft Exchange or an email gateway that supports Microsoft ActiveSync (such as Lotus Traveler). It might be useful to be able to keep your work and personal contacts on one mobile device instead of carrying two phones around all day. Due to the differences in the Contacts and People applications, it is easier to add accounts through the Android Settings screens.

1. From the Home screen, touch the Menu button, and touch Settings.

2. Touch Accounts. On the DROID Incredible 2, this menu item is labeled Accounts & Sync.

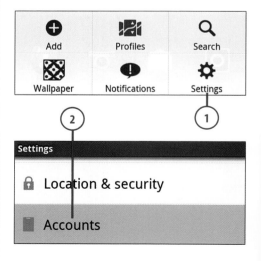

3. Touch Add Account.

4. Touch Corporate Sync. On the DROID Incredible 2 this is labeled Exchange ActiveSync.

The Sequence Might Not Be Exactly the Same

Depending on the DROID you are using, the exact sequence of events and screen layout might not exactly match the following steps. However, you do use the same information to add either the ActiveSync or Corporate account.

5. Enter your corporate email address.

6. Enter your corporate network password.

7. Enter your company's domain name. This is the domain name you see every time you log in to your computer at the office.

8. Enter your network username.

9. Touch Next.

Error Adding Account? Guess the Server

Your DROID tries to work out some information about your company's ActiveSync setup. If it can't, you are prompted to enter the ActiveSync server name manually. If you don't know what it is, you can try guessing it. If, for example, your email address is dsimons@allhitradio.com, the ActiveSync server is most probably webmail.allhitradio.com. If this doesn't work, ask your email administrator.

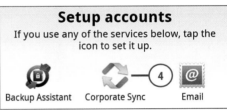

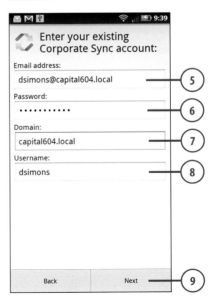

10. Touch Done.

Remove an Account

To remove an account, from the main Contacts screen, touch the Menu button and touch Accounts. Touch the account you want to remove. At the bottom of the screen, touch Remove Account.

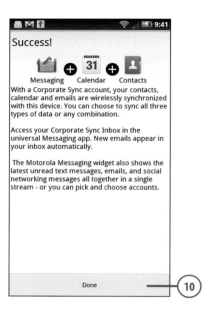

Working with Contacts (DROID Pro and CHARGE)

The Contacts application runs on the DROID Pro and CHARGE only. Although the DROID X2 has a Contacts app icon, it really launches an app called Dialer, which is covered later in this chapter.

Navigating Contacts

The Contacts app actually has three screens. The first one you see shows you your address book, but there are two others that have specific functions.

1. From the Home screen, touch the Contacts icon. If you have the Launcher open, the Contacts icon looks like an address book.

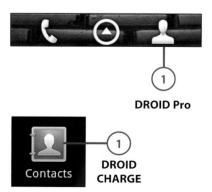

2. Touch to add a new contact.

3. Touch a contact to see more information about them.

4. Touch a contact picture to show the Quick Contact bar.

5. Touch to show the contact's phone numbers. Touch one to call it.

6. Touch to see all the contact's information.

7. Touch to show a list of numbers you can send text messages to. Touch a number to send a text message to it.

8. Touch to see a list of email address for the contact. Touch an address to send an email.

9. Touch to see a view the contact's Facebook profile, send a Facebook message, write on the wall, or poke him (DROID Pro only).

10. Touch to launch the mobile version of Facebook in the web browser (DROID Pro only).

11. Swipe left to see the contact history. On the DROID CHARGE, touch the History tab.

12. Swipe to the right to see the contact's social network statuses. On the DROID CHARGE, touch the Activities tab.

13. Touch to search your contacts (DROID CHARGE only).

No Facebook for Android Integration

Even if you have installed the Facebook for Android app from the Android Market (read more about apps in Chapter 10, "Working with Android Applications"), when you use the Quick Contact bar, it only connects to Facebook using the web browser. There is no built-in intelligence to launch the Facebook app.

DROID Pro

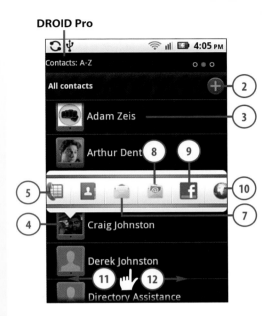

DROID CHARGE

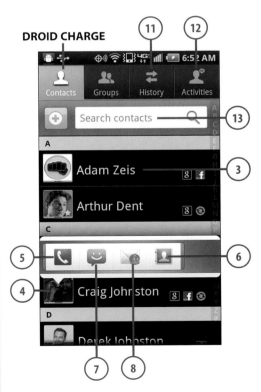

Using Contact History

The Contact history shows the history of all types of communication with your contacts including phone calls, text messages, and emails.

1. Touch to add a contact (DROID Pro only).

2. Touch to see the Quick Connect bar (DROID Pro only).

3. Touch to see the full details of the contact method.

4. Touch to clear the contact history list.

5. Touch to search all contacts (DROID Pro only).

6. Touch to select which contact types to see in this view—for example, limit this view to only show calls and not show Facebook messages (DROID CHARGE only).

DROID Pro

DROID CHARGE

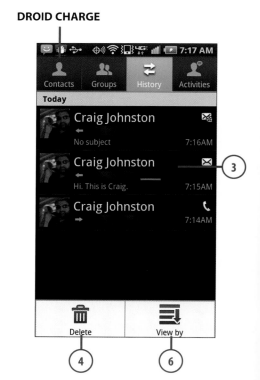

Checking a Contact's Status

If you have added contacts that belong to social networks like Facebook, you can check their statuses right from the Contacts app.

1. Touch to add a contact.

2. Touch to see the Quick Connect bar.

3. Touch to see the full social network status details.

4. Touch to mark the contact as a favorite.

5. Touch to see the full status post.

DROID Pro

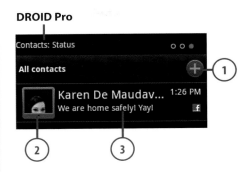

DROID Pro

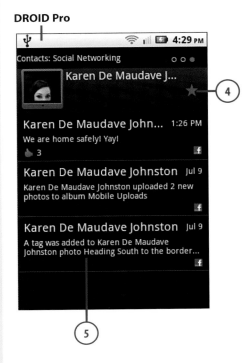

6. Touch to see the previous post.

7. Touch to see the next post.

8. Touch to write on the contact's Facebook wall, send her a private Facebook message, or send an email.

9. Touch to view a Facebook photo album. This launches the mobile Facebook site in the web browser.

10. Touch to see a photo and comment on it.

11. Touch to comment on any Facebook post (DROID CHARGE only).

Samsung DROID CHARGE Has Better Social Integration

As you can see, the Samsung DROID CHARGE has much better social networking integration within the Contacts app. It took just one touch to see all social networking status updates and interacting with the updates was much easier. On the DROID Pro it took three screens and the updates are not as accurate.

DROID Pro

DROID CHARGE

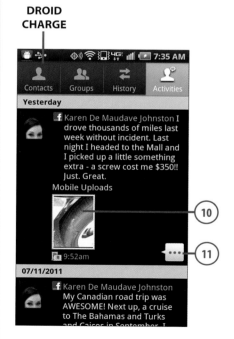

Editing a Contact

Sometimes you need to make changes to a contact or add additional information to it.

1. Touch the contact to edit.

2. Press the Menu button and touch Edit.

3. Touch to change or remove the contact picture.

4. Touch to enter a middle name, name prefix and suffix, and phonetic names.

5. Touch to add a phone number.

6. Touch to delete a phone number.

7. Touch to add an email address.

8. Scroll down for more contact information.

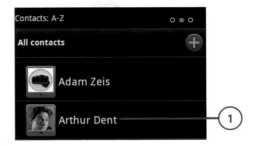

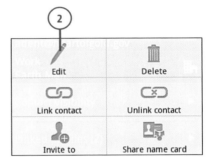

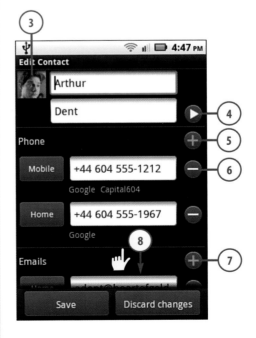

9. Touch to hide the contact from the list of contacts.

10. Touch to see additional information.

11. Touch to select a specific ringtone to play when this contact calls you.

12. Touch to send this contact straight to voicemail when he calls you (DROID Pro only).

13. Touch to save your changes.

14. Touch to discard your changes.

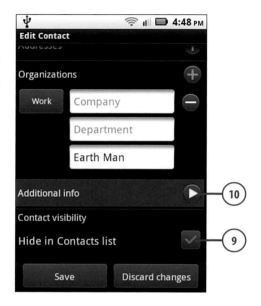

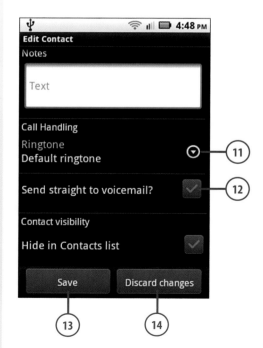

Adding a Contact Photo

The contact photo is normally added automatically when a social network account is linked to a contact. However, you can manually add a photo picture.

1. Touch the Contacts icon.

2. Touch the name of the person you want to add a contact picture to.

3. Press the Menu button and touch Edit.

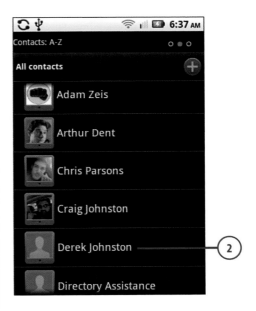

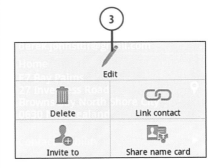

4. Touch the contact photo place holder.

5. Touch to add a photo already on your DROID.

6. Touch Gallery.

7. Touch the album where the photo is located.

Take a photo with your camera instead of using an existing one

Choose Files if the photo is not in the Gallery app

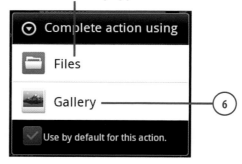

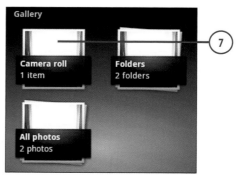

8. Touch the photo.

9. Drag the cropping box to select the area of the photo you want to use as the contact photo.

10. Touch to save the cropped photo as the contact photo.

11. Touch to save the changes to the contact card.

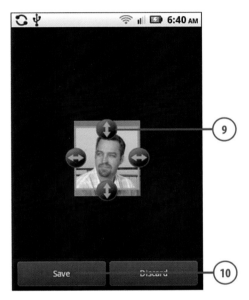

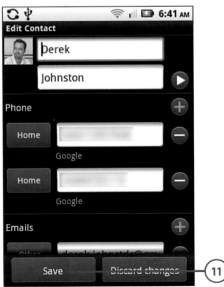

Adding and Managing Contacts

As you add contacts to your work email account or Google account, those contacts are synchronized to your DROID automatically. When you reply to or forward emails on your DROID to an email address that is not in your Contacts, those email addresses are automatically added to the contact list or merged into an existing contact with the same name. You can also add contacts to your DROID directly.

Adding Contacts from an Email

To manually add a contact from an email, first open the email client (either email or Gmail) and then open an email. Please see Chapter 5, "Emailing," for more on how to work with email.

1. Touch the blank contact picture to the left of the sender's name.

2. Touch Add to Existing Contact to merge the contact's name and email address with an existing contact.

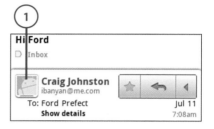

DROID Pro

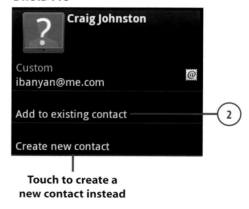

Touch to create a
new contact instead

DROID CHARGE

3. Touch the existing contact name you want to merge the new email address into.

DROID Pro

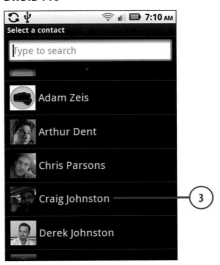

DROID CHARGE

Touch to create a
new contact instead

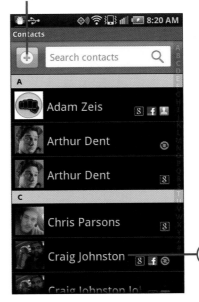

4. Touch Save.

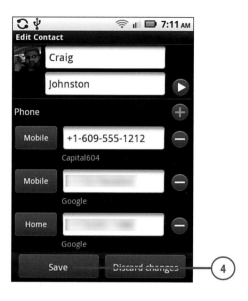

Adding a Contact Manually

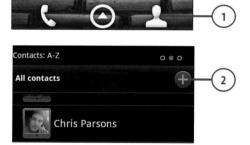

1. Touch the Contacts icon on the Home screen.

2. Touch to add a new contact. On the DROID CHARGE you will be prompted to choose where to add the new contact.

Choose where to add the new contact (DROID CHARGE only)

3. Touch the down arrow next to Family Name to fill in more fields, such as middle name and phonetic given name.

4. If you touch the plus symbol next to an item you can add multiples of that item, such as multiple phone numbers.

Choose an IM Account Type

When you add a new IM account, touch the label to the left of the entry field to change the type of IM to AIM, Windows Live, Yahoo!, and many other types of IM accounts.

5. Touch a minus sign to delete an item.

6. Scroll down to see more items to enter.

7. Touch to enter additional information such as birthdays, anniversaries, nicknames, and so on.

8. Scroll down to see more items to enter.

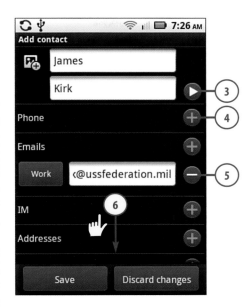

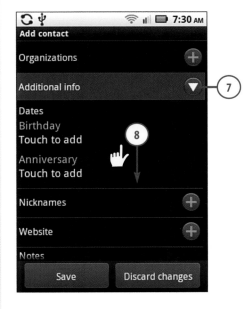

9. Touch to choose a unique ring-tone to play when this person calls you.

10. Touch to send calls straight to voicemail when this contact calls you (DROID Pro only).

11. Touch to save the new contact.

No vCard Attachment Support

Unlike other smartphones, your DROID cannot deal with vCards attached to emails. Although you can import vCards if they are saved to your SD card, when they arrive as an email attachment there is no way to import them. You need to import them using your desktop email client.

Contacts Settings (DROID Pro Only)

There are a couple of settings for the Contacts app such as choosing which accounts to save new contacts to and deleting contacts from your SIM card.

1. Touch the Contacts icon on the Home screen.

2. Press the Menu button and touch More.

3. Touch Settings.

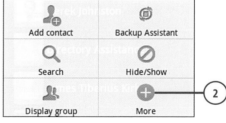

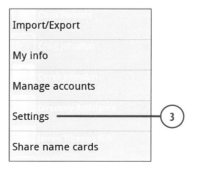

4. Select the accounts you want new contacts to be added to. You cannot unselect Phone Contacts.

5. Touch to delete contacts from your SIM card.

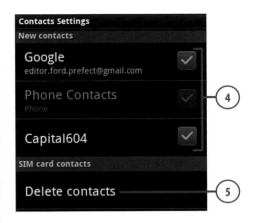

Creating Contact Groups

You can create contact groups—such as Facebook, Work email, Favorites, and so on—and then divide your contacts among them.

1. Touch the Contacts icon on the Home screen.

2. Press the Menu button and then touch Display group.

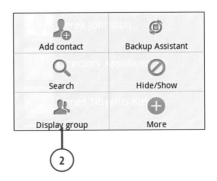

3. Touch the group you want to show the contacts from, or touch All Contacts to show everyone in all groups.

4. Touch to create a new group and add contacts to that group. You cannot delete groups so be careful when you create them.

Groups on Your DROID CHARGE

It is easier to view groups on your DROID CHARGE. Just touch the Groups tab on the top of the Contacts screen and then touch the group you want to view. Press the Back button to return to the list of groups.

Touch to choose a group

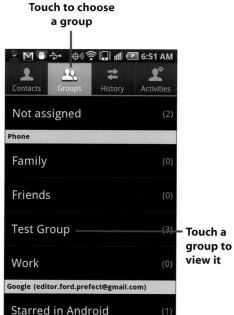

Touch a group to view it

Showing or Hiding Contacts

You can choose to hide certain contact groups from the main contacts display. For example, hide all Facebook contacts, but show all other contacts.

1. Touch the Contacts icon on the Home screen.

2. Press the Menu button and touch Hide/Show on your DROID Pro. On your DROID CHARGE, touch More and then touch Display Options.

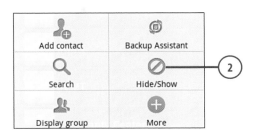

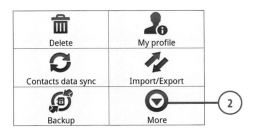

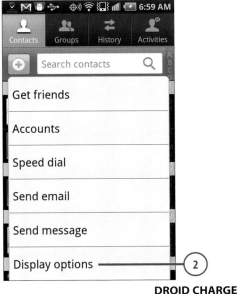

DROID CHARGE

3. Select groups to hide from the list of contacts.

4. Touch to expand a group to see subgroups of contacts.

5. Touch Done to save your changes.

DROID Pro

DROID CHARGE

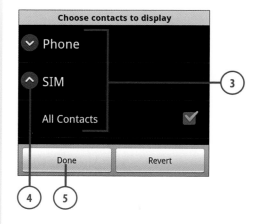

Linking and Separating Contacts

As you add contacts to your DROID, they are automatically merged if the new contact name matches a name that's already stored. Sometimes you need to manually join contacts together or separate them if your DROID has joined them in error.

Joining Contacts Manually

1. Open Contacts then scroll to and touch the contact that you want to join a contact to. Press the Menu button and touch Edit Contact.

2. Press the Menu button and touch Link Contact. On the DROID CHARGE, use the Join Contact option.

DROID Pro

DROID CHARGE

DROID Pro

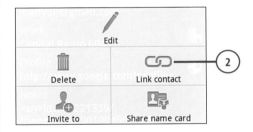

DROID CHARGE

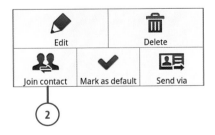

3. Touch the name of the contact you want to link with.

4. Press the Back button to save the changes and return to the contacts list.

DROID Pro

Separating Contacts

1. Open Contacts. Scroll to and touch the contact that you want to separate.

DROID CHARGE

DROID Pro

DROID CHARGE

2. Touch or press the Menu button and touch Unlink Contact. On the DROID CHARGE touch Joined Contacts.

3. Touch the contact that you want to unlink.

DROID Pro

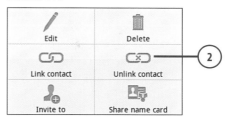

DROID CHARGE

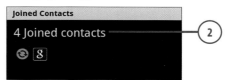

DROID Pro

DROID CHARGE

4. Touch the contact that you want to see after you unlink them. You are choosing which contact to display in the edit view (DROID Pro only).

DROID Pro

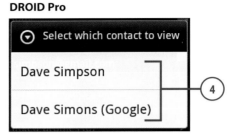

It's Not All Good

Can't Tell Which Contact Is Which

One bad thing about unlinking contacts is when it shows you the list of linked contacts, you sometimes can't tell which ones are which.

People (DROID Incredible 2)

Your DROID Incredible 2 doesn't have a Contacts app but it does have a People app that serves the same purpose.

Navigating People

The People app actually has three screens. The first one you see shows you your address book, but there are two others that have specific functions.

1. From the Home screen, touch the People icon.

2. Touch to see all contacts.

3. Touch to choose and manage groups of contacts.

4. Touch to see people in online directories such as Facebook, Flickr, Twitter, and your company's internal address book.

5. Touch to see the call history.

6. Touch a contact to see all of their information.

7. Touch to add a new contact.

8. Touch a contact's picture to see the Quick Contact bar.

9. Touch to call the contact.

10. Touch to view all the contact's information.

11. Touch to send the contact a text message (SMS) or multimedia text message (MMS).

12. Touch to send an email to the contact using either the Mail app or the Gmail app.

13. Touch to see the contact's Facebook information page.

Working with a Contact

When you touch a contact you can do a variety of things such as see their Facebook status, call history, and gallery of photos.

1. Touch the contact you want to view from the People all contacts view.

2. Touch to see linked contacts. Read more about linked contacts later in this chapter.

3. Touch to call the contact.

4. Touch to send the contact a text message.

5. Touch the send the contact an email.

6. Touch to choose a different ringtone to play when this contact calls you.

7. Touch to block this contact when he calls you.

8. Touch to edit the contact.

9. Touch to see the contact's details.

10. Touch to see the contact's Facebook status.

11. Touch to see the contact's pictures.

Editing Contact Information

You can edit a contact by adding new information, changing existing information, or deleting information.

1. Touch the contact you want to edit from the People all contacts view.

2. Press the Menu button and touch Edit.

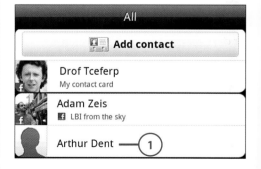

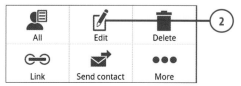

3. Touch to choose or change the contact picture.

4. Touch to choose which name to display or to edit the name.

5. Touch to delete items, such as phone numbers and email addresses.

6. Touch to add items.

7. Scroll down to see all contact information.

8. Touch to save your changes.

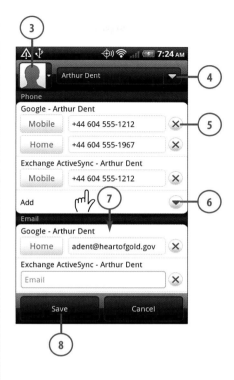

Editing a Contact Picture

You can change the contact picture or choosewhich online account you want the picture to come from.

1. Touch the contact you want to edit from the People all contacts view.

2. Press the Menu button and touch Edit.

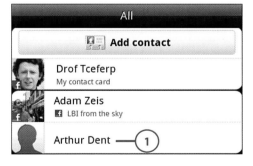

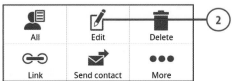

3. Touch to choose or change the contact picture.

4. Touch Gallery to select a picture.

Choosing Where the Picture Comes From

In Step 4 you can choose to take a picture with your DROID's camera and use it as the contact picture, choose an existing picture in the Gallery, or choose a service to synchronize the contact picture from. When you choose an online service, you are telling your DROID that it must update the contact's picture to the one the person uses on that online service automatically. The more services that person belongs to, the more choices you'll see on this screen.

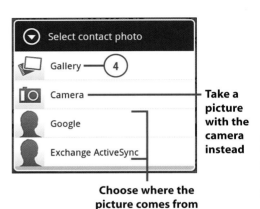

Take a picture with the camera instead

Choose where the picture comes from

5. Touch an album.

6. Touch the picture to use as the contact picture.

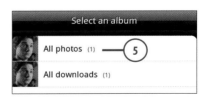

7. Select the part of the picture to use as the contact picture by moving and resizing the green crop box.

8. Touch Save.

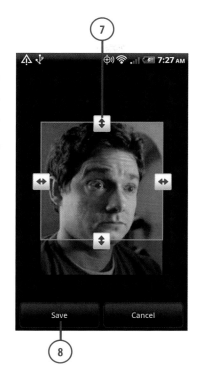

Linking and Separating Contacts

As contacts are added to your DROID Incredible 2, contacts are automatically merged if the new contact name matches a name that's already stored. Sometimes you need to manually join contacts or separate them if your DROID has joined them in error.

Joining Contacts Manually

1. Open People then scroll to and touch the contact that you want to join a contact to.

2. Touch the broken link icon.

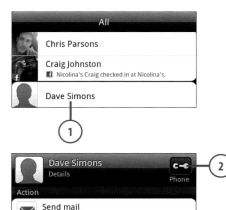

3. Touch All Contacts.

4. Touch the contact you want to link with.

5. Touch Done.

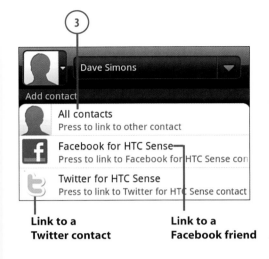

Link to a Twitter contact

Link to a Facebook friend

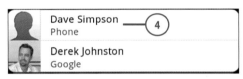

Choose the Correct Name

If you link two contacts that have slightly different names—for example someone who you have in your contacts using their maiden and married names or someone who uses a stage name—before you touch Done to save the contact, you can touch the name on the top of the screen and choose which name you want displayed in your contacts list.

Touch the name drop-down

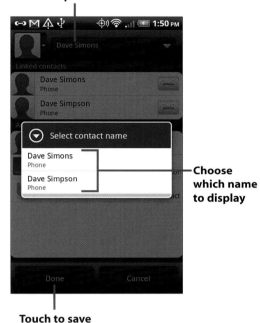

Choose which name to display

Touch to save

Separating Contacts

1. Open People. Scroll to and touch the contact that you want to separate.

2. Touch the Linked icon.

3. Touch the contact that you want to unlink. Notice how the icon changes to a broken link.

4. Touch Done.

Showing or Hiding Contacts

You can choose to hide certain contact groups from the main contacts display. For example, you can hide all Facebook contacts but show all other contacts.

1. Touch the People icon on the Home screen.

2. Touch the Menu button and touch View.

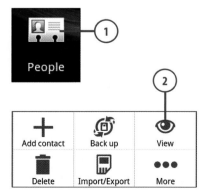

3. Select groups to hide from the list of contacts.

4. Touch to expand a group to see subgroups of contacts.

5. Touch Done to save your changes.

Dialer (DROID X2 and DROID 3)

Your DROID X2 and DROID 3 have both the Contacts and Dialer icons, however both launch the Dialer app, which has a contacts component. Read more detail about the Dialer app Chapter 2, "Using the Phone and Google Voice."

Navigating Dialer Contacts

When you launch the Dialer it shows the Contacts tab by default. While on the Contacts tab you can do many of the functions you can in the real Contacts app.

1. From the Home screen, touch the Contacts or Dialer icons.

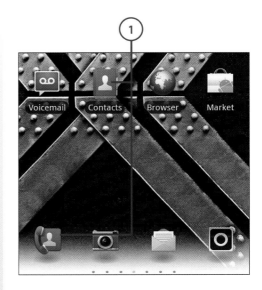

2. Touch a contact to see all of the information.

3. Touch a contact picture to reveal the Quick Contact bar.

4. Touch to call the contact.

5. Touch to send the contact a text message or multimedia message.

6. Touch to send the contact an email.

7. Touch to view the contact's Facebook profile, send a Facebook message, write on the wall, or poke them.

8. Touch to visit the contacts profile pages on Google and Facebook.

9. Touch to select which groups of contacts to display.

10. Touch to search for contacts.

11. Touch to add a new contact.

12. Touch a letter to jump to contacts whose first name starts with that letter.

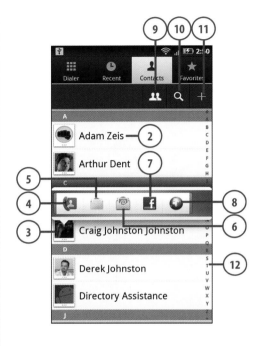

Change the Sort Order

By default your DROID sorts contact names by first name. You can change it to sort by last name if you like. To do this, press the Menu button and touch Display Options. Touch Sort list by to change the sort order.

Adding a Contact

Sometimes you need to add a contact to your contact list manually.

1. Touch the green plus symbol on the Dialer Contacts tab.

2. Choose an account where the new contact should be stored.

3. Touch to remember your selection and use it automatically in future.

4. Touch OK.

5. Touch to add a contact picture.

6. Touch to add more name information including middle name, prefix and suffix, and phonetic name spelling.

7. Touch to add a new item, for example to add a new phone number or email address.

8. Touch to remove an item.

9. Scroll down to see more items.

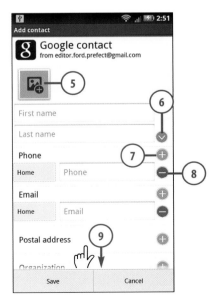

10. Touch to see more items.

11. Touch to add an Instant Messaging (IM) account such as AOL, Skype, Yahoo!, Google Talk, and more.

12. Touch to add a website.

13. Touch to save the contact information.

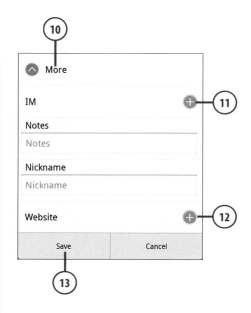

Editing a Contact

Sometimes you need to edit a contact to change some information about it.

1. Touch the contact you want to edit on the Dialer Contacts tab.

2. Press the Menu button and touch Edit Contact.

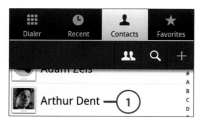

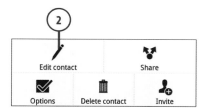

3. Touch to add or change the contact picture.

4. Touch to change the item type. For example, change the mobile phone to the home phone.

5. Touch to add a new item, for example a new phone number or email address.

6. Touch to edit the item.

7. Touch to remove an item.

8. Touch to save your changes.

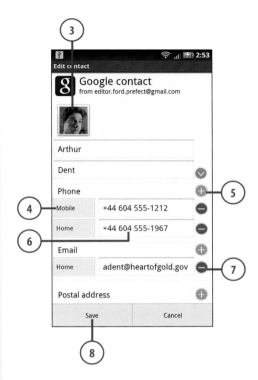

Adding a Contact Picture

Many times your contact's pictures are synchronized automatically from online services such as Facebook. However if a contact is not part of any online services, you can add a contact picture manually.

1. Touch the contact you want to add the contact picture to on the Dialer Contacts tab.

2. Press the Menu button and touch Edit Contact.

3. Touch to add a picture.

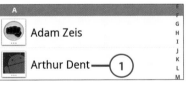

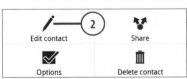

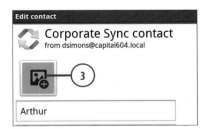

4. Touch to choose a photo from your Gallery.

5. Touch to choose a photo from your Gallery again.

Why Would I Choose Files?

Most of the time, pictures you have downloaded or taken with your camera are in the Gallery app. However, there might be instances when you need to browse your DROID's file system to find images. When you choose Files, you can also connect to disks that are attached to your home network that are shared and available. If you have images stored on those shared disks, you can select them here.

6. Touch the picture you want to use.

Take a photo with your camera instead

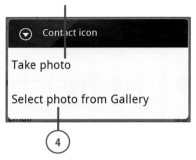

Select a picture that is not in your Gallery

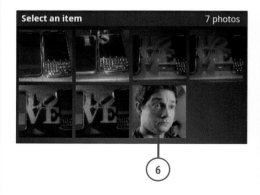

7. Touch to save your changes.

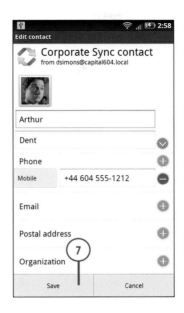

Linking and Separating Contacts

As contacts are added to your DROID X2, contacts are automatically merged if the new contact name matches a name that's already stored. Sometimes you need to manually join contacts together or separate them if your DROID has joined them in error.

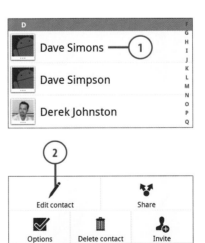

Joining Contacts Manually

1. Touch the contact that you want to join a contact to.

2. Press the Menu button and touch Edit Contact.

3. Press the Menu button and touch Join.

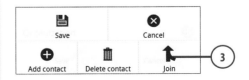

4. Touch the contact you want to link with. Your DROID X2 suggests a contact which is often correct but if the suggestion is incorrect, touch Show All Contacts to make your own selection.

5. Touch Save.

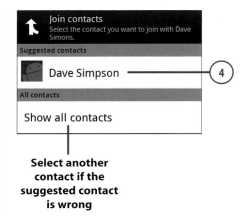

Select another contact if the suggested contact is wrong

Separating Contacts

1. Touch the contact that you want to separate.

2. Press the Menu button and touch Edit Contact.

3. Press the Menu button and touch Separate.

4. Touch OK.

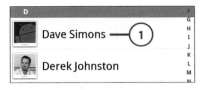

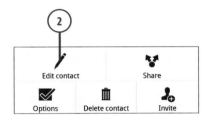

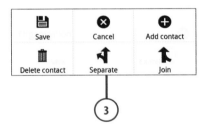

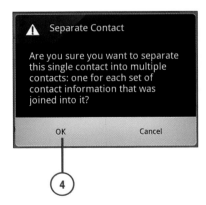

Showing or Hiding Contacts

You can choose to hide certain contact groups from the main contacts display. For example, you can hide all Facebook contacts but show all other contacts.

1. Press the Menu button and touch Display Options from the Contacts tab in Dialer.

2. Select groups to hide from the list of contacts.

3. Touch to expand a group to see subgroups of contacts.

4. Touch to save.

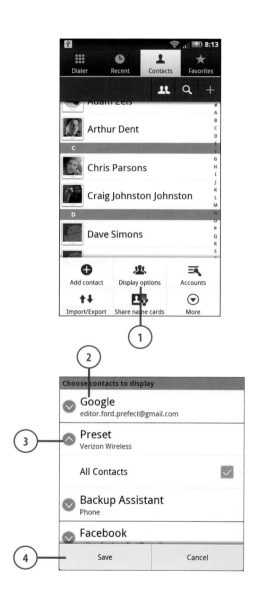

Adding a Contact to Your Home Screen

If you communicate with some contacts so much that you are constantly opening and closing the Contacts or People application, a quicker solution might be to add a shortcut to the contacts on the Home screen.

1. Touch and hold on the Home screen where you want to add the contact shortcut.

2. Touch Shortcuts.

3. Touch Contact. On the DROID Incredible 2 you touch Person.

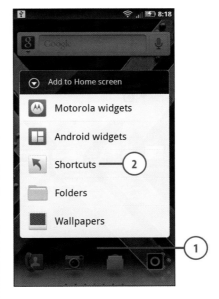

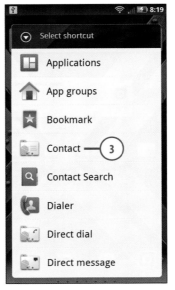

4. Touch the contact you want to add to the Home screen.

Final Steps Adding a Contact on Your DROID Incredible 2

For your DROID Incredible 2, after Step 4 you are prompted to select the type of communication you want to trigger when you touch the contact shortcut. All contact types are listed including phone and email.

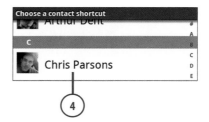

Select contact method

In this chapter, you learn about your DROID's Phone application. You learn how to place and receive calls, manage them, and use Google Voice as your voicemail. Topics include the following:

→ Placing and receiving calls

→ Managing in-progress calls

→ Conducting conference calls

→ Setting up Google Voice

→ Using Google Voice for voicemail

Using the Phone and Google Voice

As with any good smartphone, your DROID has a great phone that allows for making and receiving calls, voicemail, three-way calling, and many other features. However, your DROID can also use Google Voice to save you money on calls and to transcribe your voicemail if you want it to. We cover your DROID's phone in detail and go over how to set up and use Google Voice.

Getting to Know the Phone Application (DROID 3, DROID Pro, DROID X2, and DROID CHARGE)

The DROID 3, DROID X2, DROID Pro, and DROID CHARGE all have very similar Phone apps, which are covered in this section. Small changes between the phones are called out.

1. Touch the Phone icon on the Home screen. The phone application launches and displays the familiar phone keypad.

2. Type a phone number.

3. Touch the green phone icon to place the call.

4. Touch and hold the 1 key to listen to your voicemail.

5. Touch the delete icon to correct a phone number, or touch and hold the icon to remove the entire number.

6. Touch the most recently dialed number to dial it again (not available on the DROID CHARGE).

7. Touch to dial using your voice. Read more about dialing using your voice later in this chapter.

8. Touch to show the phone dialer tab.

9. Touch to show the most recent activity. This is also known as the call log.

10. Touch to see all of your contacts.

11. Touch to see only your favorite contacts.

12. Touch to send a text message (DROID CHARGE only).

>>> Go Further

PAUSES AND WAITS

To add a two-second pause (symbolized by a ",") or a wait (symbolized by a ";") into your phone number, touch the Menu button, and touch the appropriate icon. When you touch Add Wait, your DROID dials the number up to the point of the wait, and then waits until it hears a response from the other side. You can insert multiple two-second pauses if you need to. Pauses and waits can be useful when using calling cards, doing phone banking, or dialing into conference calls.

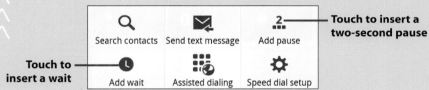

Touch to insert a two-second pause

Touch to insert a wait

Search contacts Send text message Add pause Add wait Assisted dialing Speed dial setup

---

International Numbers

Normally when you dial international numbers, you have to use some kind of code before the country code. With your DROID, you don't need to know that special code; just type a plus sign, the country code, and the rest of the number, dropping any zeros before the area code.

Getting to Know the Phone Application (DROID Incredible 2)

1. Touch the Phone icon on the Home screen.

2. Enter a phone number or a name using the letters on each number key.

3. Touch to correct a phone number, or touch and hold the to remove the entire number.

4. Touch to hide the keypad.

5. Touch to display the People application.

6. Touch to dial the number.

7. Indicates the person is already in your contacts.

8. Touch to add the number to an existing contact, or create a new contact using the number.

9. Indicates an outgoing call.

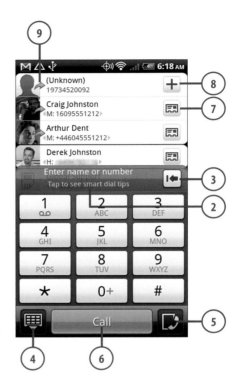

Receiving a Call (DROID 3, DROID Pro, and DROID X2)

When you receive a call on the DROID 3, DROID Pro, or DROID X2, you have two choices for handling it.

1. Slide the green phone icon to the right to answer the call.

2. Slide the red phone icon to the left to send the call to voicemail.

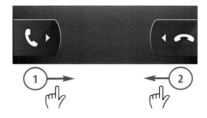

Receiving a Call (DROID Incredible 2)

When you receive a call on the DROID Incredible 2, you have two choices for handling it.

1. Slide the gray bar down to answer the call.

2. Slide the gray bar up to send the call to voicemail.

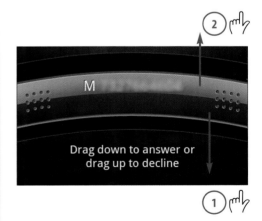

Receiving a Call (DROID CHARGE)

When you receive a call on the DROID CHARGE, you have two choices for handling it.

1. Slide the green puzzle piece into the open puzzle piece hole answer the call.

2. Slide the red puzzle piece into the open puzzle piece hole to send the call to voicemail.

3. Slide the gray puzzle piece into the open puzzle piece hole to mute the ringtone.

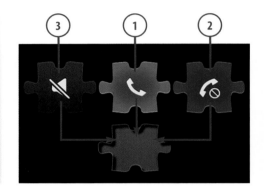

Handling Missed Calls

1. If you miss a call, the missed call icon displays in the status bar.

2. Pull down the status bar to see how many calls you've missed.

3. Touch the missed calls notification to open your DROID's Call Log.

4. Touch the green phone icon to the right of a log entry to call the person back.

5. Touch a Call Log entry to see other actions to take for the call.

6. Touch Call *<Name>* to call the person back.

7. Touch Send Text Message to send a text message (SMS) to the caller.

8. Touch View Contact to see the caller's contact card if the person is already in your Contacts.

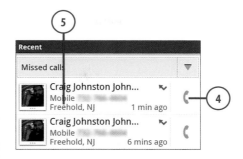

Clear the Call Log

To clear the Call Log, touch the Menu button and touch Clear Call Log. To clear just one entry from the Call Log, touch and hold that entry and choose Remove from Call Log.

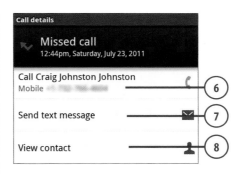

Placing a Call (DROID3, DROID Pro, DROID CHARGE, and DROID X2)

You can place calls on your DROID in a few ways including manually dialing a number into the Phone application, touching a phone number in a contact entry, commanding your DROID using your voice, and touching a phone number on a web page, in an email, or in a calendar appointment.

Dialing from a Contact Entry

1. Touch the Contacts application on the Home screen.

2. Touch the name of the person you want to call.

3. Touch the phone number you want to call.

DROID 3, Pro, X2

DROID CHARGE

DROID 3, Pro, X2

DROID CHARGE

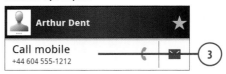

DIAL IN FEWER STEPS

If you want to dial someone using fewer steps, touch and hold the contact's name in the contact list and then touch Call Contact. Finally touch the number to call.

Dialing Using Your Voice

1. Touch and hold the Search button until the Speak Now box appears.

2. Say "Call" and the person's name. For example "Call Arthur Dent." Your DROID finds that person in your Contacts and displays the phone numbers on the screen.

3. Touch the phone number you'd like to call.

Be More Specific

If you know there are multiple phone numbers for a particular contact you want to call, you can be more specific when speaking your command. For example you can say "Call Arthur Dent Mobile" to dial Arthur Dent's mobile number immediately. This reduces the steps needed to place the call.

Set a Default Number

To set a default number for a contact, open the contact entry. Touch and hold the phone number you want to set as the default, and touch Make Default Number.

Bluetooth Support for Voice Dialing

If you are using a Bluetooth headset, hold down the button on your headset to activate voice commands on your DROID. Speak a command such as "Call Arthur Dent".

Options While on a Call

While on a phone call, you can mute and unmute the call, switch the audio between your DROID and a Bluetooth headset, bring up the dial-pad, and enable the speaker phone.

1. Touch the Dialpad icon to display the dialpad and type additional numbers during the call, for example for phone banking. While the dialpad is displayed, the dial-pad icon changes to the Hide Dialpad icon. Touch it to hide the dialpad.

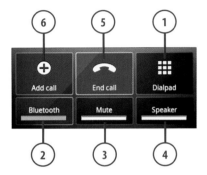

2. Touch the Bluetooth icon to switch the call from your DROID to a Bluetooth headset. When you switch you hear a beep in your Bluetooth headset.

Need to Know More about Bluetooth

Read Chapter 4, "Connecting to Bluetooth, Wi-Fi, and VPNs," for help on setting up Bluetooth headsets.

3. Touch the Mute icon to mute the call.

4. Touch the Speaker icon to enable your DROID's speaker phone.

5. Touch the End Call icon to end the call.

6. Touch Add Call to start a confer-ence call.

Conference Calling

While on a call, you can create an impromptu conference call by adding callers.

1. Touch the Add Call icon. Your current call goes on hold.

2. A second dialpad appears. Either type in the number for the person to call, or touch the Contacts icon to dial from your contacts.

3. Touch Merge to join the two calls together into one conference call.

Adding Multiple Callers
While on the conference call touch the Add Call icon again to add another caller. The number of callers you can add to a conference call depends on what your wireless carrier supports.

Indicates you are still on a call

End second call

4. Touch to end the last call you added to the conference.

5. Touch to end the entire conference call.

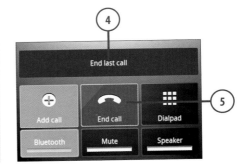

Placing a Call (DROID Incredible 2)

You can place calls on your DROID in a few ways including manually dialing a number into the Phone application, touching a phone number in a contact entry, commanding your DROID using your voice, and touching a phone number on a web page, in an email, or in a calendar appointment.

Dialing from a Contact Entry

1. Touch the People application on the Home screen.

2. Touch the name of the person you want to call.

3. Touch the phone number you want to call.

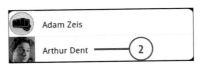

DIAL A CONTACT QUICKLY

An alternative method of dialing a contact from the People application is to touch and hold on the contact's name and then choose which number you want to call.

Dialing Using Your Voice

1. Touch and hold the Search button until the Speak Now box appears.

2. Say "Call" and the person's name. For example "Call Arthur Dent." Your DROID finds that person in your Contacts and displays the phone numbers on the screen.

3. Touch the phone number you'd like to call.

Be More Specific

If you know there are multiple phone numbers for a particular contact you want to call, you can be more specific when speaking your command. For example you can say, "Call Arthur Dent Mobile," to list Arthur Dent's mobile number at the top of the list. This reduces the steps needed to place the call.

Bluetooth Support for Voice Dialing

If you are using a Bluetooth headset, hold down the button on your headset to activate voice commands on your DROID. Speak a command such as, "Call Arthur Dent."

Options While on a Call

While on a phone call, you can mute and unmute the call, switch the audio between your DROID and a Bluetooth headset, bring up the dial pad, and enable the speaker phone.

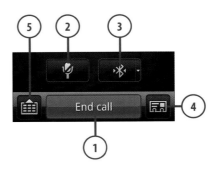

1. Touch to end the call.

2. Touch to mute your microphone.

3. Touch to switch between a Bluetooth headset, speakerphone, or ear piece.

Need to Know More about Bluetooth?

Read Chapter 4 for help on setting up Bluetooth headsets.

What Is ADR6350?

In step 3 when you touch to switch between audio devices, you see a list that shows your Bluetooth headset (if you have one paired), speaker, and ADR6350. HTC ADR6350 is the actual model number of the HTC Incredible 2. When you select ADR6350 the audio from the call goes to the Incredible 2's ear piece.

Bluetooth headset

Incredible 2's ear piece   **Incredible 2's speakerphone**

4. Touch to view the contact information for the person you called, or who called you, as long as it is already stored in the People application. If the person is not listed in your contacts you are able to add him.

5. Touch to show the dialpad. This is useful if you need to use phone menus while doing phone banking or dialing into conference calls.

6. Touch the Menu button to reveal more options.

7. Touch to add a person to the conversation. See more about conference calling in the next section.

8. Touch to show the People application.

9. Touch to perform a Flash. Like landline phones, this puts the call on hold and enables you to take a second incoming call.

10. Touch to switch between a Bluetooth headset, speakerphone, or ear piece.

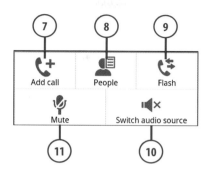

Where Does the Audio Go?

By default, if you are paired with your Bluetooth headset before you make or receive a call, the call audio uses the headset. If you turn your Bluetooth headset on after you are on the call, your DROID automatically switches the audio to it.

11. Touch to mute your audio.

Conference Calling

While on a call, you can create an impromptu conference call by adding callers.

1. Touch the Menu button and touch Add Call.

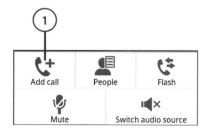

2. After using the second dialpad that appears to type the number for the person to call or touching the Contacts icon to dial from your contacts, both calls display on screen. The original call is on hold.

3. Touch to merge the calls together into a three-way conference call.

How to Drop the Second Call

While on a conference call, to drop the last person added to the call, touch the Menu button and touch Flash. This turns the call back into a one-on-one call with the original party.

Active call

On-hold call

Touch to drop the second call

Configuring the Phone Application

You can control how the Phone application works in many ways including whether to display your Caller ID information, how to handle call forwarding and call waiting, and changing your voicemail settings.

Call Settings (DROID Incredible 2)

1. From the Home screen, touch the Menu button and touch Settings.

2. Touch Call.

3. Touch Voicemail Service to choose a service to handle your voicemails. This is normally your carrier; however, you can use a different service such as Google Voice.

4. Touch to change the voicemail settings.

5. Touch to clear any voicemail notifications (if there are any).

6. Touch to enable Auto Retry. This will cause your DROID to automatically redial the number if the call fails.

7. Touch to enable hearing aid support. If you don't use a hearing aid, make sure this option is disabled because it amplifies in-call volume.

8. Touch to choose your preferred network mode. Your choices are Global Mode (automatically switches between GSM/UMTS and CDMA), CDMA Mode (only uses CDMA), and GSM/UMTS (only uses GSM/UMTS).

9. Touch to change the GSM call settings. This option is only visible when you are on a GSM network.

10. Touch to change the CDMA call settings. This option is only visible when you are on a CDMA network.

11. Scroll down for more settings.

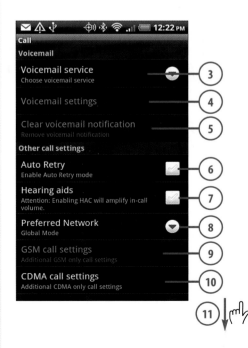

12. Touch enable or disable assisted dialing and change the assisted dialing settings.

What Is Assisted Dialing?

Assisted dialing, when enabled, helps you dial international numbers while travelling outside your home country. Choose the country you are in while you travel. Assisted dialing is really unnecessary if you always format your phone numbers in the internationally recognized format of +<country code><area code without a leading 1 or 0><number>.

13. Touch to set the phone settings.

Extra Phone Settings

Your DROID Incredible 2 has a feature that enables you to send a text message to someone if you want to acknowledge they called, but are too busy to accept the call. In this screen you can edit that default text message, and select whether to always edit the message before you send it. In addition, you can enable or disable whether your Incredible 2 automatically saves unknown contacts to the People app.

Edit default text message

Enable editing the message before sending

Save unknown contacts to the People app automatically

Sound Settings (DROID Incredible 2)

1. From the Home screen, touch the Menu button and touch Settings.

2. Touch Sound.

3. Touch to select the default ringtone to use for incoming calls. This ringtone will play unless you have set a specific custom ringtone for a particular contact.

4. Touch to enable or disable the feature that mutes the ringtone that is playing when it detects you have picked up your DROID.

5. Touch to enable or disable the feature that makes your DROID play the ringtone extra loud when it detects that it is in your pocket.

6. Touch to enable or disable the feature that causes your DROID to enable the speakerphone if it detects you have flipped it over so that the screen is facing down onto a hard surface.

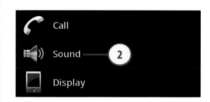

Adding Custom Ringtones

In step 3, when you touch to select a default ringtone, you can also touch New Ringtone and select an audio file already saved on your phone.

7. Touch to change the default notification sound that plays when a new notification is displayed in the status bar, such as for a missed call.

8. Touch to enable or disable audible touch tones when you touch numbers on the phone keypad. You can also set whether long or short tones play.

9. Scroll down for more settings.

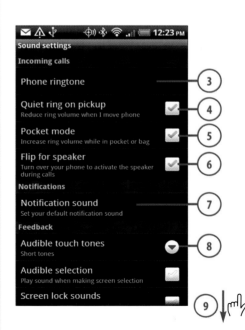

10. Touch to select whether to enable or disable an alert when you call an emergency number such as 9-1-1. You can select a vibration or a tone to play.

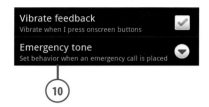

Call Settings (DROID3, Pro, X2, CHARGE)

1. From the Home screen, touch the Menu button and touch Settings.

2. Touch Call Settings.

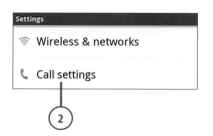

3. Touch to select your voicemail service provider. If you have Google Voice installed, you can select between Google Voice or your carrier.

4. Touch to change your voicemail settings. You might want to enter a new phone number for a different voicemail service instead of using your carrier's.

5. Touch to choose the carrier voicemail service used. Your choices are Voicemail or Visual Voicemail (if your carrier offers this).

6. Touch to choose the carrier voicemail service used when you are roaming. Your choices are Voicemail or Visual Voicemail (if your carrier offers this).

7. Touch to enable or disable assisted dialing and change the assisted dialing settings.

8. Touch to choose what happens when an incoming call is received. Your choices are play the ringtone only, say the caller's phone number and then play the ringtone, or keep repeating the caller's number.

9. Touch to enable Auto Retry. This will cause your DROID to automatically redial the number if the call fails.

10. Touch to enable TTY mode and select which mode you want to use. The DROID supports Full, HCO, and VCO.

11. Scroll down for more settings.

DROID3, Pro, X2

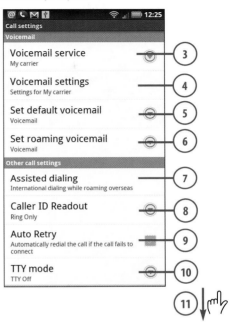

DROID CHARGE

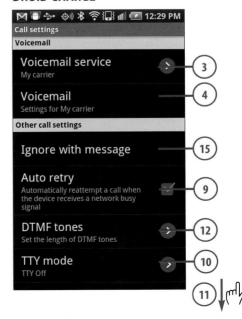

12. Touch to enable or disable audible touch tones when you touch numbers on the phone keypad, and whether they play long or short tones.

DROID3, Pro, X2

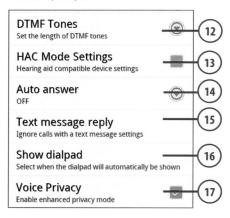

13. Touch to enable or disable support for hearing aids, also known as HAC. If you don't use a hearing aid, make sure this option is disabled because it amplifies in-call volume.

14. Touch to enable or disable auto answer and choose how many seconds your DROID must wait before automatically answering an incoming call.

15. Touch to enable or disable the feature where your DROID sends a text message to an incoming call that you ignore. You can edit the text message.

16. Touch to choose when the dial pad is automatically shown. Your choices are when calling service numbers, custom numbers, or during all calls.

DROID CHARGE

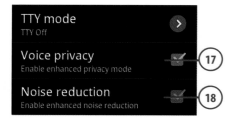

17. Touch to enable or disable an extra level of encryption on your voice calls. It has no effect on call quality.

18. Touch to enable or disable enhanced noise reduction while on a phone call (DROID CHARGE only).

Sound Settings (DROID3, Pro, X2, CHARGE)

1. From the Home screen, touch the Menu button and touch Settings.

2. Touch Sound.

3. Touch to select the default ringtone to use for incoming calls. This ringtone plays unless you have set a specific custom ringtone for a particular contact.

4. Touch to change the default notification sound that plays when a new notification is displayed in the status bar, such as for a missed call.

5. Touch to enable or disable audible touch tones when you press numbers on the dial pad.

6. Scroll down for more settings.

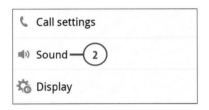

Adding Custom Ringtones

In step 3 when you touch to select a default ringtone, you can also touch New ringtone and select an audio file already saved on your phone.

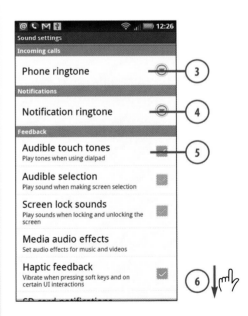

7. Touch to select whether to enable or disable an alert when you call an emergency number such as 9-1-1. You can select a vibration or a tone to play.

8. Touch to enable or disable an audio tone to play when a call is connected.

9. Touch to enable or disable an audio tone to play when your DROID loses its connection to the network.

10. Touch to enable or disable an audio tone to play when you start roaming off your home network.

DROID 3, Pro, X2

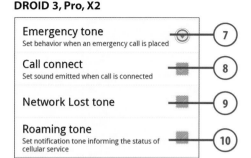

DROID Charge

Google Voice

Regular Google Voice, if set up on your DROID, enables you to save money on international calls and have your voicemails transcribed into text. If you upgrade Google Voice, which is free, you have some extra features, such as choosing your own personalized phone number or setting up simultaneous ringing.

Setting Up Google Voice

If you want to start using the Google Voice features on your DROID, you need to go through some setup steps first. If you do not have Google Voice installed, download it from the Android Market. See Chapter 10, "Working with Android Applications," for more on how to use Android Market.

1. Touch the Voice icon on the Home screen.

2. Read the welcome information from Google Voice and touch Next.

3. Select the account to use for Google Voice if you synchronize to multiple Google Accounts.

4. Touch Sign In to proceed.

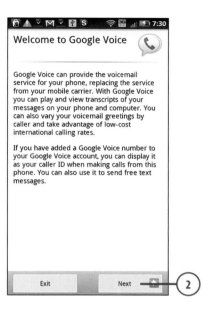

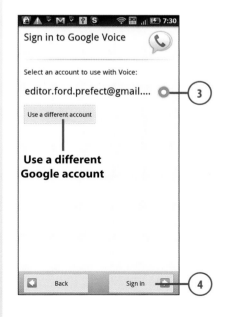

5. Touch Allow.

6. Touch Next.

7. Touch Add This Phone.

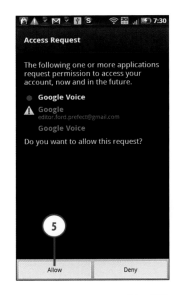

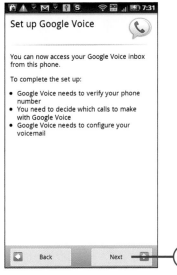

8. Touch Next to verify your phone number automatically.

9. Touch Next after your number has been verified.

10. Select how you want to use Google Voice.

11. Touch Next.

12. Touch Next.

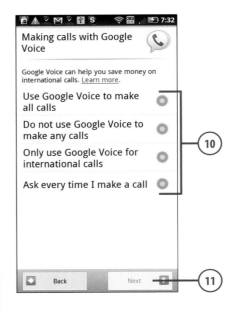

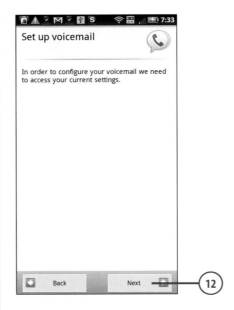

13. Touch the number displayed on the screen if you are told that you must dial a number to configure voicemail. (This depends on your carrier.)

14. Touch Configure.

15. Touch Google Voice.

16. Touch OK.

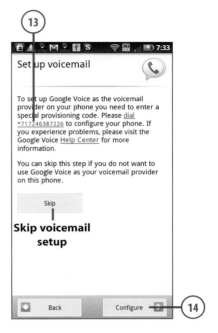

Reading Settings Forever

When you touch OK in Step 16, your DROID might show the Reading Settings screen forever. The only way to correct this is to completely power off and power on your DROID. After you power your DROID back on, however, the voicemail will be set to Google Voice.

Upgrading Google Voice

If you want to use the advanced features of Google Voice, you need to upgrade your account. Upgrading is free of charge. To upgrade, use your desktop computer to go to http://google.com/voice and log in using your Google account.

1. Click the cog icon and choose Voice Settings.

2. Click Get a Google Number.

3. Click Continue.

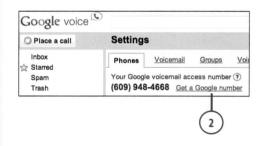

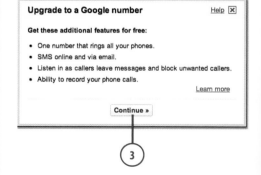

4. Type search criteria in the box to search for available Google Voice phone numbers, including numbers that spell words.

5. Select a number from the options.

6. Click Continue.

7. Click Continue on the Confirm Your Number screen.

8. Enable call forwarding from your current mobile number to your new Google Voice number by following the instructions on the screen and touching Done.

It Might Cost You

Please be aware that some wireless carriers charge to forward calls, many times on a per-forwarded-call basis, so check with your carrier before enabling the option to forward calls.

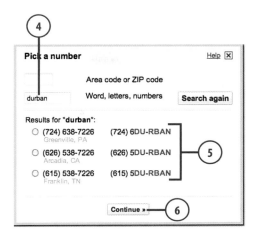

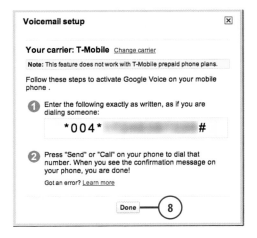

Playing Back and Managing Google Voicemails

When you receive new Google voice-mails, you can find them in your Google Voice Inbox. Launch Google Voice and follow these steps to play them back and manage them.

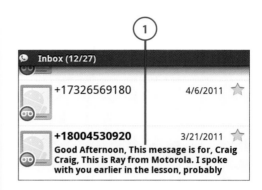

1. Touch a new voicemail to read it or play back the audio.

2. Touch the play icon to play the message audio.

Read the transcript

Sometimes Google Guesses

Grayed-out words in a voicemail transcript are words that Google Voice is unsure of.

3. Touch or press the Menu button to see actions you can take on the voicemail.

4. Touch to call the person who left the voicemail.

5. Touch to send a text message to the person who left the voicemail.

6. Touch to add the person who left the voicemail to Contacts (People on the Incredible 2).

7. Touch to mark the person who left the voicemail as one of your favorite contacts.

8. Touch to archive the voicemail.

9. Touch More to refresh the message or delete it.

Activate speakerphone

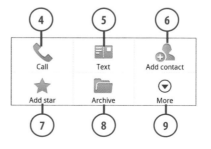

Send SMS and Check Your Balance

Google Voice enables you to send and receive text messages (SMS). You can use the main screen menu to do this plus check your Google Voice account balance, and filter the view by label. Touch the Menu button to reveal the menu.

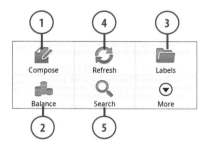

1. Touch Compose to send an SMS. The text message shows your Google Voice number as the sender.

2. Touch Balance to check your Google Voice account balance.

3. Touch Labels to filter the Inbox view by label. Labels include Voicemail, SMS, Recorded, and more.

4. Touch to refresh the Google Voice Inbox.

5. Touch to search the Google Voice Inbox.

Google Voice Balance

When you start using Google Voice to make phone calls, those calls are not free. You must carry a Google Voice balance. When you sign up for Google Voice you get $0.10 but you'll use that up pretty quickly. When you touch the Balance icon you are only able to see your balance; you are not able to add to it. For that you must use your desktop computer to log in to Google Voice at http://voice.google.com

Google Voice Settings

After you have been using Google Voice for a while, you may want to change some of the settings.

1. Touch the Menu button and touch More.

2. Touch Settings.

3. Touch to enable or disable Do Not Disturb, which sends all incoming calls to voicemail.

4. Touch Making Calls to change how Google Voice is used.

5. Touch This Phone's Number to select which phones you want Google Voice on.

6. Touch to choose where Google Voice voicemails are played back. Your choices are speakerphone or handset earpiece.

7. Touch to sign out of Google Voice.

8. Touch to configure Google Voice synchronization and notification settings.

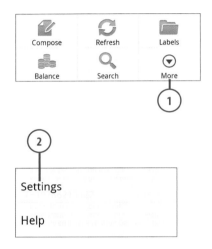

Google Voice Sync and Notifications

1. Touch to enable or disable synchronizing your Google Voice Inbox.

2. Touch to change how your DROID's background data synchronization is handled.

3. Touch to enable and disable notifications via text message when you receive a new item in your Google Voice inbox.

4. Touch to enable or disable notifications when you receive a new Google Voice message.

5. Touch to enable or disable vibration along with Google Voice inbox notifications.

6. Touch to enable or disable a blinking notification light when you receive a new item in your Google Voice inbox.

7. Touch to choose a ringtone when a new item arrives in your Google Voice inbox.

In this chapter, you learn about your DROID's audio and video capabilities, including how your DROID plays video and music, and how you can synchronize audio and video from your desktop computer. Topics include the following:

3

→ The Music application for audio and music

→ The Gallery application for video

→ YouTube

→ Making music purchases from Amazon

→ Streaming your music from Amazon's Cloud Drive

Audio and Video

Your DROID is a strong multimedia smartphone with the ability play back many different audio and video formats. The large screen enables you to turn your DROID sideways to enjoy a video in its original 16:9 ratio. You can also use your DROID to search YouTube, watch videos, and even upload videos to YouTube right from your phone.

The Music Application—Audio (DROID 2, Pro, X2, CHARGE)

Let's take a look at how the Music application works and how to enjoy hours of music while you work.

1. Touch the Music icon on the Home screen.

2. Touch Artists to filter the view by artist. Touch an artist's name to reveal songs by that artist and then touch a song to play it.

3. Touch Albums to filter the view by album title. Touch an album name to reveal songs on that album and then touch a song to play it.

4. Touch Songs (or All on the DROID CHARGE) to filter the view by song title. This shows all songs by all artists. Touch a song to play it.

5. Touch Playlists to show any music playlists that you have synchronized to your DROID. Playlists are covered later in the chapter.

6. Touch Genres to filter the view by genre. Touch an album name to reveal songs on that album and then touch a song to play it (not available on the DROID CHARGE).

7. Press the Menu button to reveal the menu.

8. Touch audio effects to control audio effects for connected wired speakers and the speaker on your DROID (not available on the DROID CHARGE).

9. Touch to shuffle all songs when played (not available on the DROID CHARGE).

10. Touch to go to Disc View, which shows your music as CDs (DROID CHARGE only).

11. Touch to change the Music app settings (DROID CHARGE only).

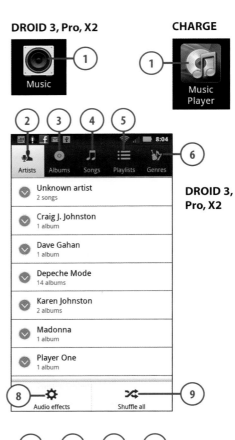

DROID 3, Pro, X2

CHARGE

DROID 3, Pro, X2

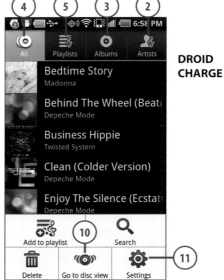

DROID CHARGE

Don't Forget the Search Button

One important thing to remember is the Search button. No matter what view you are in, you can always touch the Search button and either type or speak your search.

>>> Go Further

SCROLL BY LETTER

In any view you can scroll through your albums, artists, or songs quickly by touching the scroll box that pops out on the right of the screen. Move it up and down to jump to items beginning with the letter of the alphabet that shows in the large box. So for example, to jump to artists that begin with the letter D, scroll until you see the letter D appear in the large box, and then release the scroll box.

This box shows the letter

Move the scroll box up and down

Controlling Playback

While your music is playing, you have some control over how it plays, and the selection of music that plays.

1. Touch to jump to the previous song in the album, playlist, or shuffle. Touch and hold to rewind the song.

2. Touch to jump to the next song in the album, playlist, or shuffle. Touch and hold to fast forward the song.

3. Touch to pause the song. The button turns into the play button when a song is paused. Touch again to resume playing a paused song.

4. Touch to open the current playlist. If the song is not in a playlist, the list of all songs displays.

5. Touch to shuffle the current playlist. This plays the songs in the playlist in random order. If the song is not in a playlist, all songs on your DROID are shuffled.

6. Touch to enable repeating. Touch once to repeat all songs, touch again to repeat the current song only, touch again to disable repeating.

7. Drag to skip through the song.

Adjust the Volume

Press the volume control on the left of your DROID to increase or decrease the volume of the music. A Media Volume window pops up and displays the volume level visually.

DROID 3, Pro, X2

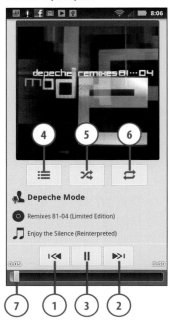

DROID CHARGE

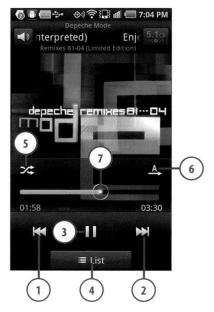

Taking More Actions

While a song is playing, if you press
the Menu button, you can take a few
actions on the song.

1. Touch to see your entire music
   library.

2. Touch to add the current song to
   a playlist.

3. Touch to use the song as the current ringtone for your DROID.

4. Touch to delete the current song.
   Confirm the deletion on the next
   screen.

5. Touch to use the song as the notification sound for your DROID.

6. Touch to control audio effects for
   connected wired speakers and
   the speaker on your DROID.

7. Touch to add the songs to the
   Quick list playlist (DROID CHARGE
   only).

8. Touch to switch between playing
   the song through a Bluetooth
   headset or speakers and your
   DROID CHARGEs speaker (DROID
   CHARGE only).

9. Touch to share the song with
   your friends using text messaging, Bluetooth, Email, or Gmail
   (DROID CHARGE only). Unless the
   song is very short, sharing via text
   messaging normally fails because
   the song's file is too large to send.

10. Touch to add the song to a
    playlist and see the song details
    (DROID CHARGE only).

DROID 3, Pro, X2

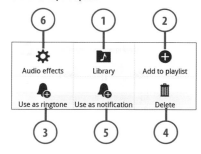

DROID CHARGE

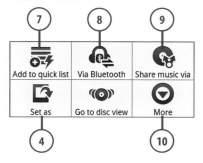

Working and Listening to Music

While your music is playing, you can continue using your DROID without interrupting the music.

1. To work on other applications while listening to music, touch the Home button. The notification bar displays an icon indicating your music is still playing.

2. To switch back to the currently playing song, pull down the notification bar.

3. Touch the song.

What If I Get a Call?

If someone calls you while you are listening to music, your DROID pauses the music and displays the regular incoming call screen. After you hang up, the music continues playing.

Managing Playlists (DROID 3, Pro, X2, CHARGE)

You can group songs together with playlists. Here is how to create them and use them.

Creating a New Playlist

To create a new playlist, you must already know the first song you want to add to it.

1. Touch and hold a song.

2. Touch Add to Playlist.

3. Touch New.

4. Type in the name of the playlist.

5. Touch Save.

Create a Playlist While Listening

You can create a playlist while listening to the song you want to use to create that playlist with. Press the Menu button and choose Add to Playlist.

Create a Playlist on Your DROID CHARGE

You don't need to know the first song of your playlist to create a new playlist on your DROID CHARGE. Simply touch the Playlists tab, press the Menu button, and touch the + symbol to add a new playlist.

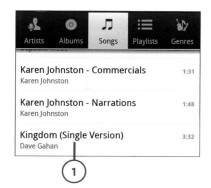

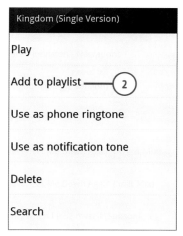

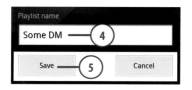

Adding a Song to an Existing Playlist

1. While listening to a song, press the Menu button. Then touch Add to Playlist.

2. Touch the playlist you want to add the song to.

Add to Playlist from Library
You can also add a song to a playlist when viewing your song library. Touch and hold on a song and then choose Add to Playlist.

Rearranging Songs in a Playlist

1. Touch a Playlist to show the songs in it.

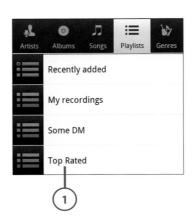

2. Touch and hold the symbol to the left of a song you want to move. Move that song up and down until it is in the right place, and then release it.

Rearranging Songs on Your DROID CHARGE

On your DROID CHARGE, when viewing a playlist contents, there is not a symbol to the left of the song names. However, you can still rearrange the songs. Press the Menu button and touch Rearrange.

The Music Application—Audio (DROID Incredible 2)

Let's take a look at how the Music application works and how to enjoy hours of music while you work.

1. Touch the Music icon on the Home screen.

2. Touch to filter your music by artist.

3. Touch to filter your music by album.

4. Touch select and manage playlists.

5. Touch to see all songs by all artists.

6. Scroll right for more options.

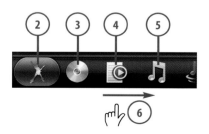

7. Touch to browse for network connected devices that support
   DLNA. These could be satellite
   receivers, network connected
   hard discs, and more.

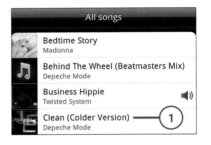

Playing Music

1. Touch a song to start playing it.

2. Touch to jump to the previous
   song.

3. Touch to pause or continue playing the song.

4. Touch to jump to the next song.

5. Touch to the Now Playing view.
   This view shows the list of songs
   queued up to play. This could be
   songs in a playlist, album, or just
   the songs in the all songs view.

6. Touch to see the album view.

7. Drag to skip back and forth
   through the song.

8. Swipe left and right through the
   "cover flow" of song cover images.

9. Touch to shuffle the current list of
   queued up songs.

10. Touch to toggle through the
    repeat options. You can turn
    repeat off, repeat all songs, or
    repeat the current song.

11. Touch to enable or disable the
    Sound Enhancement, which
    enables you to choose preset
    equalizer settings and enable or
    disable SRS. SRS provides audio
    enhancement and simulated sur-
    round sound.

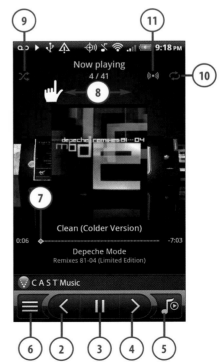

Taking More Actions

While a song is playing, if you press the Menu button, you can take a few actions on the song.

1. Touch to share the song over Bluetooth. This actually sends the music file to another device over Bluetooth. This normally only works when sending to a computer, not to another phone.

2. Touch to add the current song to a playlist. You are presented with a list of playlists. Touch one, or add a new playlist.

3. Touch to use the song as the current ringtone for your DROID.

4. Touch to enable or disable the Sound Enhancement, which enables you to choose preset equalizer settings and enable or disable SRS.

5. Touch to see more options.

6. Touch to turn shuffle on and off.

7. Touch to repeat the song.

8. Touch to see the song details like the name of the file, disc and track numbers, and location of the song on your DROID.

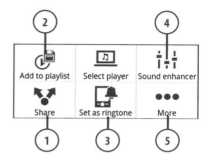

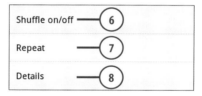

Working and Listening to Music

While your music is playing, you can continue using your DROID without interrupting the music.

1. To work on other applications while listening to music, press the Home button. The notification bar displays an icon indicating your music is still playing.

2. To switch back to the currently playing song, pull down the notification bar.

3. Touch the song.

What If I Get a Call?

If someone calls you while you are listening to music, your DROID pauses the music and displays the regular incoming call screen. After you hang up, the music continues playing.

Managing Playlists (DROID Incredible 2)

You can use playlists to group songs together. Here is how to create them and use them.

Creating a New Playlist

1. Touch to see the list of playlists.

2. Touch to add a new playlist.

3. Type the name of the playlist.

4. Touch to add songs to the playlist.

5. Touch the check box to select one or more songs.

6. Touch to add the selected songs to the playlist.

7. Touch Save.

Select All or Nothing

While selecting songs to add to your playlist, if you press the Menu button you can touch Select All or Unselect All.

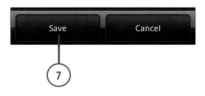

Adding a Song to an Existing Playlist

1. While listening to a song, press the Menu button. Then touch Add to Playlist.

2. Touch the playlist you want to add the song to.

Add to Playlist from Library

You can also add a song to a playlist when viewing your song library. Touch and hold on a song and then choose Add to Playlist.

Rearranging Songs in a Playlist

1. Touch a Playlist to show the songs in it.

2. Press the Menu button and touch Change Order.

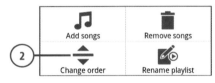

3. Touch and hold the three lines to the left of a song you want to move. Move that song up and down until it is in the right place and then release it.

4. Touch Done to save your changes.

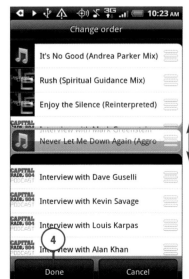

Rename or Delete a Playlist

To rename or delete a playlist, start by touching and holding the name of the playlist you want to change.

1. Touch and hold the playlist you want to rename or delete.

2. Touch to rename the playlist.

3. Touch to delete the playlist.

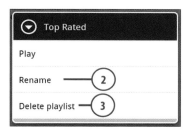

Songs Not Deleted

When you clear or delete a playlist, remember that the songs listed in them are not deleted from your DROID. Playlists are just groupings of song names.

Playing and Sharing Videos with the Gallery Application (DROID CHARGE)

The Gallery application enables you to view pictures and video, and share pictures and video with people via MMS, Bluetooth, YouTube, and email. We cover viewing and sharing videos in this chapter and cover pictures in Chapter 9, "Taking, Storing, and Viewing Pictures."

1. Touch the Gallery icon to launch the Gallery application.

2. Albums that contain videos have a little play icon. Touch an album to view available videos.

3. Touch a video to start playing it.

4. Touch to toggle between playing the video normally, stretching it to fit the screen, and playing it in 4:3 mode.

5. Drag left and right to move rapidly forward and backward through the video.

6. Touch to pause the video, and touch again to resume playing it.

7. Touch to enable 5.1 surround sound. This only works if the headset or speakers you are using support it.

8. Touch to adjust the playback volume. You can also use the volume control buttons on the left side of your DROID.

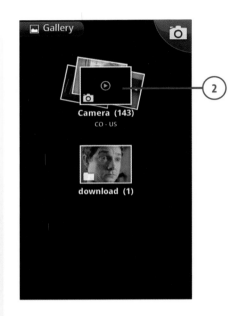

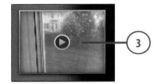

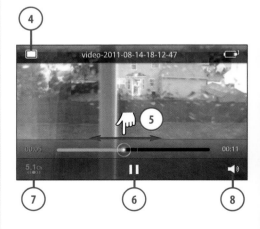

Sharing Videos

1. Touch and hold the video you want to share.

2. Touch Share.

3. Touch a method for sharing the videos.

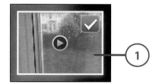

Small Video Size

You can only share very small videos from your DROID. The video file size cannot exceed 3 Mb, which is about one minute of high-quality video, or about two minutes of low-quality video. This is true for all types of video sharing, except MMS, where the videos can only be 30 seconds long.

Bluetooth Sharing Might Fail

Many phones do not accept incoming Bluetooth files, but devices like computers do. Even on computers, the recipient must configure her Bluetooth configuration to accept incoming files.

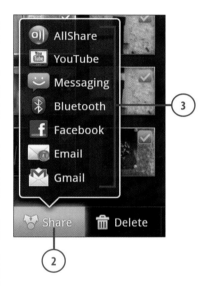

Sharing Video on YouTube

If you have not previously entered your YouTube account, you are prompted to do so before you can upload your video.

1. Enter the title of your video.

2. Touch to see more details.

3. Touch to change the YouTube account you are using, if you need to.

4. Select whether your video is public for everyone to see, or private.

5. Scroll down to see more options.

6. Enter a description for your video.

7. Touch to send the video location information.

8. Touch Upload.

Sharing Video on Facebook

1. Enter the title of your video.

2. Enter the description of the video.

3. Touch Upload.

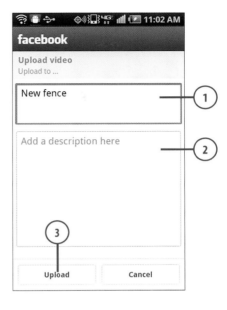

Deleting Videos

1. Touch and hold the video you want to delete.

2. Touch Delete.

More Details About Your Video

To see more details about a spe-cific video, touch More and then touch Details while the video is selected. The application displays the title, type of video (for exam-ple, 3GPP or MP4), the date the video was created, the album it is in, and the GPS location (if this is available).

Playing and Sharing Videos with the Gallery Application (DROID Incredible 2)

1. Touch the Gallery icon to launch the Gallery application.

2. Touch to see the albums view.

3. Touch to see the Facebook view where pictures and videos from your Facebook friends are visible.

4. Touch to see the Flickr view where your Flickr friend's pictures are visible.

5. Touch to see the DLNA (Digital Living Network Alliance) view, which shows you any DLNA-compatible devices on your Wi-Fi network to which you can connect to view pictures and videos on.

6. Touch the Videos album to see all videos.

7. Touch a video to play it.

8. Touch the screen while a video is playing to see playback options.

9. Touch to enable or disable SRS audio enhancement.

10. Touch to toggle between stretching the video to fit the screen or playing it in its original aspect radio.

11. Drag left and right to move rapidly forward and backward through the video.

12. Touch to pause the video. Touch again to resume playback.

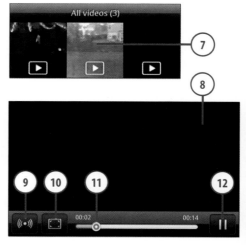

Sharing Videos

1. Touch to share one or more videos.

2. Touch to choose your method of sharing. Remember that if you want to share a video on YouTube, you must first have a YouTube account.

3. Touch one or more videos to mark them as selections.

4. Touch Next.

Small Video Size

Depending on your cellular carrier, you might only be able to share very small videos from your DROID. The video file size normally cannot exceed 3 Mb, which is about one minute of high-quality video, or about two minutes of low-quality video. This is true for all types of video sharing, except MMS, where the videos can be as short as 30 seconds.

Bluetooth Sharing Might Fail

Many phones do not accept incoming Bluetooth files, but devices such as computers do. Even on computers, the recipient must configure her Bluetooth configuration to accept incoming files.

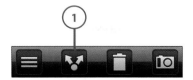

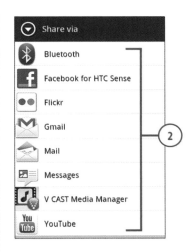

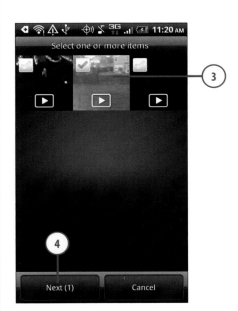

Deleting Videos

1. Touch to select one or more videos for deletion.

2. Touch one or more videos. Videos you select have a red X.

3. Touch Delete.

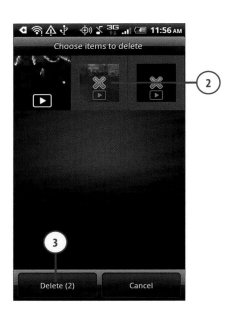

More Details About Your Video

To see more details about a specific video, touch and hold on a video and then touch Details while the video is selected. The application displays the title, type of video (such as 3GPP or MP4), the date the video was created, the album it is in, and the GPS location (if this is available).

Playing and Sharing Videos with the Gallery Application (DROID 3, X2, and Pro)

1. Touch the Gallery icon to launch the Gallery application.

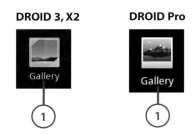

2. Touch the All Videos albums to see videos on the DROID Pro and touch any album (such as Camera Roll or My Library) on the DROID 3 and X2.

3. Touch a video to load it ready for playing.

4. Touch to Launch the video camera.

5. Shows where the video was recorded if that information was included in the video (DROID 3 and X2 only).

6. Shows the date and time the video was recorded.

7. Touch to share the video with friends using YouTube, Facebook, and other methods.

8. Touch to use the Quick Load function as long as you have previously set it up.

9. Touch to comment on the video if the video is from Facebook (DROID 3 and X2 only).

10. Touch to start playing the video.

11. Touch to delete or edit the video (DROID Pro only).

DROID 3, X2

DROID Pro

DROID 3, X2　　　　**DROID Pro**

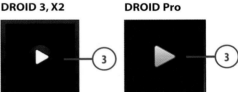

DROID 3, X2

DROID Pro

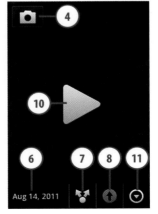

Deleting Videos

1. Touch and hold the video you want to delete.

2. Touch Delete.

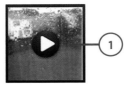

Editing Videos

1. Touch and hold a video you want to edit.

2. Touch Edit.

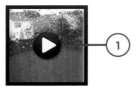

3. Touch to edit the video information such as the name and description (DROID 3 and X2 only).

4. Touch to edit the video tags.

5. Touch to trim and edit the video.

6. Move the start and end marker left and right to trim the video.

7. Press the Menu button to reveal more options.

8. Touch to extract the current frame from the video.

9. Touch to add a title to the video.

10. Touch to resize the video.

11. Touch to remove the audio from the video.

12. Touch to save the edited video.

DROID 3 and X2

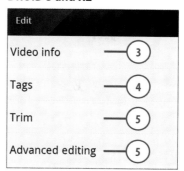

DROID Pro

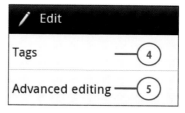

DROID 3 and X2

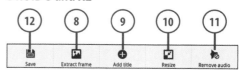

DROID Pro

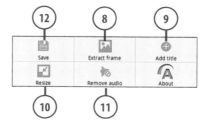

Recording Videos with the Camera Application (DROID CHARGE)

The Camera application enables you to take pictures, and record video. We cover the video recording feature of the Camera application in this chapter. Using the Camera application to take pictures is covered in Chapter 9.

Recording Video

1. Touch to launch the Camera app.

2. Slide the switch from the still camera to the video position.

3. Touch to start recording video. Touch the icon again to stop recording.

Camera

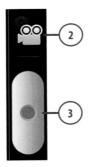

It's Not All Good

Video Focusing Is Off

While the camera is in still camera mode, when you take pictures, your DROID can focus the image using a mechanical auto-focus feature. However while recording videos, the focus remains frozen so if you bring your DROID too close to someone or something, the video goes out of focus.

Changing Video Settings

Before you record a video, you can change some settings that can alter how the video is recorded.

1. Touch to see the video settings.

2. Touch to enable or disable the outdoor visibility enhancement.

3. Touch to enable the timer and set it between 2 and 10 seconds. The timer makes the camera wait before starting to record video.

4. Touch to set the video resolution. Your choices are 320×240, 720×480, 176×144 (good for MMS), 640×480, and 1280×720

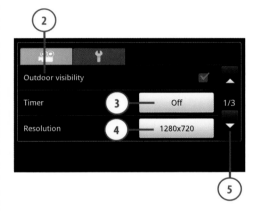

5. Touch to see more settings.

6. Touch to set white balance. You can either leave it on automatic or choose preset levels for cloudy days, daylight, and so on.

7. Touch to set the video effects. You can either leave the video as normal or apply effects to it such as negative and sepia.

8. Touch to set the video quality; your choices are Superfine, Fine, and Normal. This controls the amount of compression that is done on the video when it saves it. Superfine has the least compression therefore looks the best, and normal has the most compression and looks the worst.

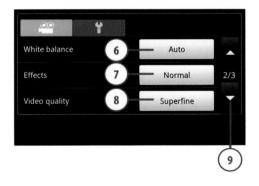

9. Touch to see more settings.

10. Touch to adjust contrast, color saturation, and sharpness.

11. Touch to see more settings.

12. Touch to enable or disable guidelines. When enabled lines are drawn in a grid pattern.

13. Touch to to enable or disable recording audio with your video.

14. Touch to enable or disable a feature that allows you to review the video after you have recorded it.

15. Touch to see more settings.

16. Touch to reset all camera and video settings back to the factory default.

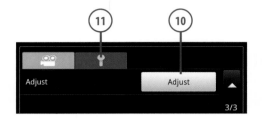

Recording Videos with the Camera Application (DROID X2 and Pro)

Recording Video

1. Touch to launch the Camcorder app. The Camcorder app is not its own app but is just the Camera app with the video mode selected.

DROID X2

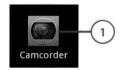

DROID Pro

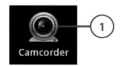

DROID X2

2. Touch to start and stop recording.

3. Touch the + and – symbols to zoom in and out (DROID X2 only).

4. Touch to reveal the right menu.

5. Touch to choose video recording scenes such as everyday, outdoor, and so on. These preset the video camera for certain situations to make the video and audio best suited for those situations.

6. Touch to choose video effects such as black and white, sepia, and so on.

7. Touch to turn the video camera light on or off. In low light situations having the light on can help illuminate the area and make your video look better.

8. Touch to switch between the camcorder and camera modes.

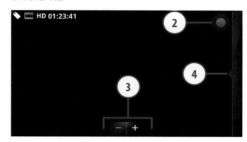

DROID Pro

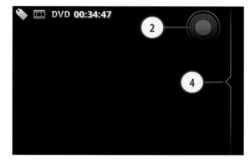

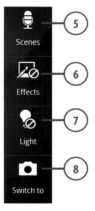

It's Not All Good

Video Focusing Is Off

While the camera is in still camera mode, when you take pictures, your DROID can focus the image using a mechanical auto-focus feature. However, while recording videos, the focus remains frozen so if you bring your DROID too close to someone or something, the video goes out of focus.

Changing Video Settings and Modes

Before you record a video, you can change some settings that can alter how the video is recorded.

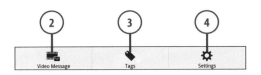

1. Press the Menu button to reveal the bottom menu.

2. Touch to change the video mode to Video Message mode, which reduces the resolution and screen size appropriate for sending video via MMS. On the DROID Pro you can also select fast or slow motion video modes.

3. Touch to add and edit video tags. Tags help you describe the video. If you have it enabled, your GPS coordinates are also tagged in the video.

4. Touch to see the Camcorder app settings.

5. Touch to set the picture resolution. See more about pictures in Chapter 9.

6. Touch to set the video resolution. Your choices are 320×240, 352×288 (good for MMS), 640×480, 720×480, and 720p (which is high-definition video and only available on the DORID X2).

7. Touch to change the camera exposure setting which ranges from 3 to –2.

8. Touch to enable or disable the camera shutter sound.

9. Touch to choose where to store your recorded video. You can choose the phone or the SD card (if you have one inserted).

DROID X2

DROID Pro

10. Scroll down for more settings.

11. Touch to choose the focus mode. This setting is only for camera mode when you are taking pictures. See Chapter 9 for more information on taking pictures.

DROID X2

12. Touch to enable or disable the Quick Upload album and set it up. The Quick Upload album is a preset location where you can share pictures and video.

13. Touch to enable or disable face detection, which allows your DROID to detect and focus on faces.

14. Touch to enable or disable the camera shutter animation.

DROID Pro

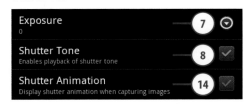

Recording Videos with the Camera Application (DROID 3)

Recording Video

1. Touch to launch the Camcorder app. The Camcorder app is not its own app but is just the Camera app with the video mode selected.

2. Touch to switch between the front and back cameras.

3. Touch to start and stop recording.

4. Touch to switch between the camcorder and camera modes.

5. Touch to turn the video camera light on or off. In low light situations having the light on can help illuminate the area and make your video look better.

6. Touch to change the brightness levels.

7. Touch to switch between normal and video message modes. Video message mode is used for sending video mesasges via MMS.

8. Touch to choose video scenes such as everyday, outdoors, concert, and so on. Choosing different scenes helps set up the camcorder for specific conditions which ultimately make the video look better and the audio sound better.

9. Touch to choose video effetcs like negative, sepia, and so on.

10. Touch to change the camcorder settings.

11. Swipe up and down to zoom in and out.

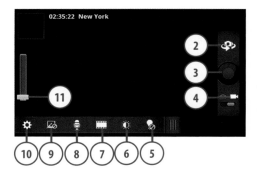

Its Not All Good

Video Focusing Is Off

While the camera is in still camera mode, when you take pictures, your DROID can focus the image using a mechanical auto-focus feature. However, while recording videos the focus remains frozen so if you bring your DROID too close to someone or something, the video goes out of focus.

Changing Video Settings

Before you record a video, you can change some settings that can alter how the video is recorded.

1. Touch to toggle between 8 megapixel widescreen and 6 megapixel widescreen pictures. See more about pictures in Chapter 9.

2. Touch set the video resolution. Your choices are 320×240, 352×288 (good for MMS), 640×480, 720×480, and 720p (which is high-definition video).

3. Touch to select where your recorded video is saved. You can choose your phone or the SD card (if one is inserted).

4. Touch to select whether your location information is included with the video.

5. Scroll down to see more settings.

6. Touch to enable or disable the fake camera shutter sound.

Recording Videos with the Camera Application (DROID Incredible 2)

Recording Video

1. Touch to launch the Camcorder.

2. Touch to switch between the camera and camcorder modes.

3. Touch to enable or disable the camcorder light. In low light situations, turning the light on helps illuminate the area.

4. Touch to start and stop recording.

5. Touch to select video effects like grayscale, sepia, and so on.

6. Touch to open the Gallery app to view videos.

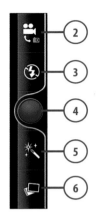

Changing Video Settings

Before you record a video, you can change some settings that can alter how the video is recorded.

1. Press the Menu button to pull out the settings panel.

2. Touch to change the Exposure, Contrast, Color Saturation, and Picture Sharpness.

3. Touch to set the white balance. You can choose settings like Daylight or Incandescent, or you can leave it set to automatic.

4. Touch to change video quality. You can choose between HD 720p (High Definition), Widescreen, and a selction of different resolutions.

5. Touch to set the video review duration.

6. Touch to enable or disable including audio with your video.

7. Scroll down for more settings.

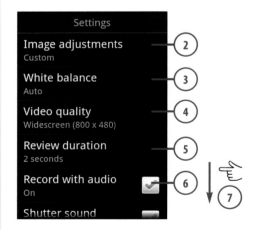

8. Touch to enable or disable the fake shutter sound.

9. Touch to reset your camcorder settings back to the factory default.

YouTube

Your DROID comes with a YouTube application that enables you to find and watch videos, rate them, add them to your favorites, and share links to YouTube videos. The YouTube application even enables you to upload new videos.

YouTube Main Screen

1. Touch the YouTube icon to launch the YouTube application.

2. Press the Menu button to see more options.

3. Touch to record a video with your DROID's camera and instantly upload it to YouTube.

4. Touch to search YouTube.

5. Touch a video to see more information about it and play it.

6. Touch return to the YouTube Home screen.

7. Touch to browse YouTube by category.

8. Touch to see your YouTube channel (if you have one). Your channel contains any videos you have uploaded to YouTube.

9. Touch to upload a video to YouTube that was previously recorded, or is stored on your DROID.

10. Touch to change the YouTube app settings.

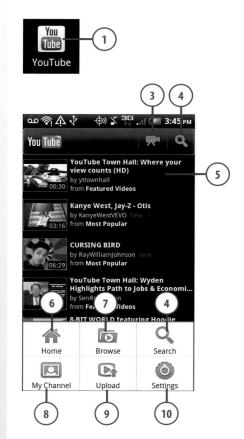

Playing a Video

While playing a YouTube video, you can rate the video, read video comments, and share the video with someone.

1. Double-tap the video, or rotate your DROID into landscape mode to see the video full screen.

2. Touch More to see more options.

3. Touch to add the video to one of your YouTube playlists. As of the writing of this book, this feature does not work.

4. Touch to add the video to your YouTube favorite video list.

5. Touch to share the video with people using MMS, email, and Facebook. When you share the video, you are only sharing the link to it.

6. Touch to copy the YouTube video link. After it's on your DROID's clipboard, you can paste it into text in any app.

7. Touch to indicate that you like the video.

8. Touch to indicate that you do not like the video.

9. Touch to flag a video as inappropriate.

10. Touch to see information about the video, including who uploaded it, the video title, description, and how many times it has been viewed.

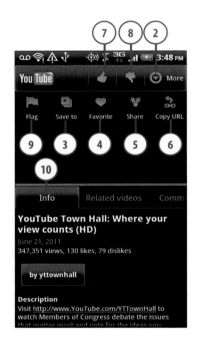

11. Touch to see YouTube videos that are related to the video you are watching. YouTube finds videos that are related because of content and key words.

12. Touch to comment on the video you are watching and read other people's comments.

13. Touch to see the YouTube channel of the person who uploaded the video.

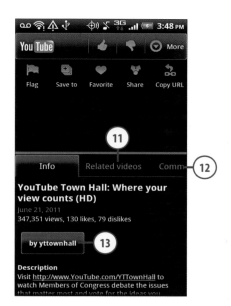

Filtering Videos

When you first open the YouTube application, you are presented with the Featured videos. You can view videos from a specific category only.

1. Touch or press the Menu button, and touch Browse.

2. Scroll to and touch a category you want to filter the view to.

3. Touch to further filter the view by time. For example, you can choose to see only videos uploaded today.

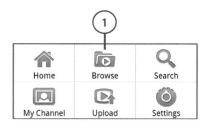

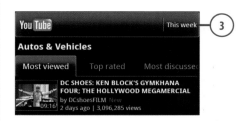

Changing YouTube Settings

If you want to clear your YouTube search history, or change the Time Filter, you can do this in the YouTube application's settings screen.

1. Press the Menu button while looking at the main YouTube screen, then touch Settings.

2. Touch to enable or disable always starting videos in high quality mode. The video might take longer to start playing and it uses more data in high quality mode.

3. Touch to set the size of the font used when a video has captions.

4. Touch to choose when your DROID uploads videos to YouTube. Your choices are Only When on WiFi and On Any Network.

5. Touch to clear your YouTube search history.

6. Touch to set the types of videos that are displayed when you search. If you set this setting to Don't Filter then no videos are filtered out based on content.

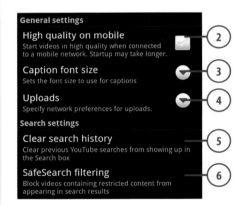

Amazon MP3

Your DROID comes with the Amazon MP3 application. This application enables you to search for and purchase music from Amazon, and download it directly to your phone or store it in the Amazon Cloud Drive. Before you start, you need an Amazon account. If you don't have one, visit Amazon.com on your desktop computer to sign up.

Amazon MP3

Setting Up the Amazon MP3 Application

After you have your Amazon.com account, you are ready to use Amazon MP3. Here is how to set it up by entering your account details.

1. Touch the Amazon MP3 Store icon on the home screen.

2. Press the Menu button after you see the main Amazon MP3 screen, and then touch Settings.

3. Touch to sign in to your Amazon account. You need to do this before using Amazon MP3.

4. Touch to enter an Amazon claim code, which can be an Amazon gift card or promotional code.

5. Touch to enable or disable forcing Amazon MP3 to only download music you have purchased when your DROID is connected to Wi-Fi.

6. Touch to clear anything that is stored in your Amazon MP3 cache such as songs, album art, and your now playing queue.

7. Touch to enable or disable Amazon MP3 playback controls on your DROID's lock screen. This enables you to control anything playing via the Amazon MP3 player even if your DROID is locked.

8. Touch to select whether to enhance your music using the equalizer.

9. Scroll down for more settings.

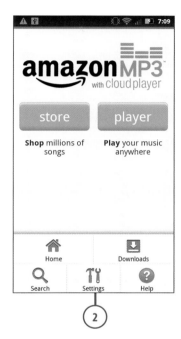

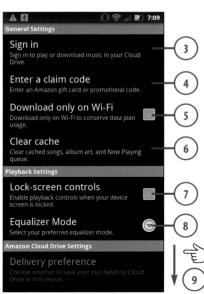

10. Touch OK to save your login information.

11. Touch to set where music you download or buy through the Amazon MP3 store is saved. You can save the music to your Amazon MP3 Cloud Drive, which saves the music to a computer at Amazon or save it to your DROID.

12. Touch to automatically download music purchases when you choose to save them to your Cloud Drive.

13. Touch to choose when your Amazon MP3 player will stream music from your Cloud Drive. You can choose all network connections, or choose between cellular and Wi-Fi or Wi-Fi only.

14. Touch to refresh your CloudDrive if you think it is not showing recent changes.

Amazon MP3 Store

The Amazon MP3 Store enables you to find and download or purchase music right on your DROID.

1. Touch to browse the Amazon MP3 Store.

2. Touch to switch to the Amazon MP3 Player.

3. Touch to search for music.

4. Touch to play a sample of a song.

5. Touch to download a free song.

6. Touch to buy a song.

Downloading Free Music

1. Touch the Free button.

2. Touch the Get button.

3. If prompted, choose where you would like to save your downloaded song. You can choose to save it in the Amazon Cloud Drive or on your DROID.

Purchasing Music

1. Touch the price of the song.

2. Touch the Buy button.

3. If prompted, choose where you would like to save your down-loaded song. You can choose to save it in the Amazon Cloud Drive or on your DROID.

In this chapter, you learn about your DROID's connectivity capabilities including Bluetooth, Wi-Fi, VPN, and web browsing. Topics include the following:

→ Pairing with Bluetooth devices
→ Connecting to Wi-Fi networks
→ Virtual Private Networks (VPN)
→ Using your DROID as a Wi-Fi Hotspot

Connecting to Bluetooth, Wi-Fi, and VPNs

Your DROID can connect to Bluetooth devices such as headsets, computers, and car in-dash systems, as well as to Wi-Fi networks, and 2G and 3G cellular networks. It has all the connectivity you should expect on a great smartphone. Your DROID can also connect to virtual private networks (VPN) for access to secure networks. Some DROIDs can even share their 3G cellular connection with other devices over Wi-Fi.

Connecting to Bluetooth Devices

Bluetooth is a great personal area network (PAN) technology that allows for short distance wireless access to all sorts of devices such as headsets, other phones, computers, and even car in-dash systems for hands free calling. The following tasks walk you through pairing your DROID to your device and configuring options.

Pairing with a New Bluetooth Device

Before you can take advantage of Bluetooth, you need to connect your DROID with the Bluetooth device, which is called pairing. After you pair your DROID with a Bluetooth device, they can connect to each other automatically in the future.

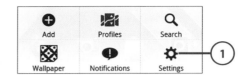

1. Press the Menu button and then touch Settings.

2. Touch Wireless & Networks.

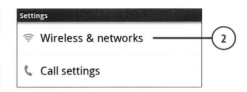

3. Touch to enable Bluetooth. On the DROID CHARGE, this setting is on the next screen.

4. Touch Bluetooth Settings.

5. Touch Device Name to change the name that your DROID uses when it broadcasts on the Bluetooth network.

6. Touch Discoverable if you want to make your DROID discoverable on the Bluetooth network. Your DROID remains discoverable for 120 seconds (two minutes). Making your DROID discoverable is necessary when someone else is trying to pair with your DROID.

7. Touch Scan for Devices to get your DROID to scan the Bluetooth network for other Bluetooth devices. If your DROID finds any other devices, they are listed on the screen. In this example, a Bluetooth headset, a BlackBerry, and a computer were found.

8. To pair with a Bluetooth device, touch the device. In this example, I am pairing with the Plantronics headset.

DROID Bluetooth FTP

On your DROID Incredible 2, there is an extra option on the Bluetooth Settings screen called Advanced Settings. If you touch Advanced Settings there is one option which is to enable Bluetooth FTP server. When this is enabled, you can drag and drop files to and from your DROID Incredible and a desktop computer without having to connect it using the USB cable.

Bluetooth Passkey

If you are pairing with a device that requires a passkey, such as a car in-dash system or a computer, the screen shows a passkey. Make sure the passkey is the same on your DROID and on the device you are pairing with. Touch Pair on your DROID, and confirm the passkey on the device you are pairing with.

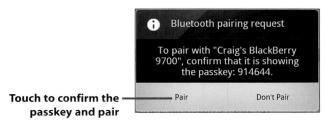

Touch to confirm the passkey and pair

---

9. If all went well, your DROID should now be paired with the new Bluetooth device.

All Zeros

If you are pairing with an older Bluetooth headset, you might be prompted to enter the passkey. Try using four zeros as the passkey. It normally works. If the zeros don't work, refer to the headset's manual.

Changing Bluetooth Options

After a Bluetooth device is paired, you can change a few options for some of them. The number of options depends on the Bluetooth device you are connecting to. Some have more features than others.

1. Touch and hold on a Bluetooth device to see available options.

2. Touch to disconnect from the Bluetooth device.

3. Touch to disconnect and unpair from the Bluetooth device. If you do this, you won't be able to use the device until you redo the pairing as described in the previous task.

4. Touch for more options.

5. Touch to connect with the Bluetooth device, if you are currently disconnected from it.

6. Touch to enable and disable using this device for phone calls. Sometimes Bluetooth devices have more than one profile. You can use this screen to select which ones you want to use.

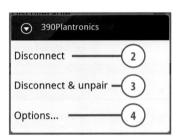

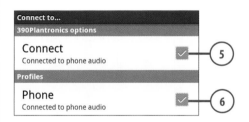

Bluetooth Profiles

Each Bluetooth device can have one or more Bluetooth profiles. Each Bluetooth profile describes certain features of the device. This tells your DROID what it can do when connected to it. A Bluetooth headset normally only has one profile such as Phone Audio. This tells your DROID that it can only use the device for phone call audio. Some devices might have this profile, but provide other features such as the Phone Book Access profile that would allow it to synchronize your DROID's address book. The latter is typical for car in-dash Bluetooth.

Quick Disconnect

To quickly disconnect from a Bluetooth device, touch the device on the Bluetooth Settings screen and then touch OK.

Wi-Fi

Wi-Fi (Wireless Fidelity) networks are wireless networks that run within free radio bands around the world. Your local coffee shop probably has free Wi-Fi, and so do many other places such as airports, train stations, malls, and other public areas. Your DROID can connect to any Wi-Fi network and provide you higher Internet access speeds than the cellular network.

Connecting to Wi-Fi

The following steps explain how to find and connect to Wi-Fi networks. After you have connected your DROID to a Wi-Fi network, you auto-matically are connected to it the next time you are in range of that network.

1. Press the Menu button and touch Settings.

2. Touch Wireless & Networks.

3. Touch to enable Wi-Fi.

4. Touch to change Wi-Fi Settings and connect to Wi-Fi networks.

5. Touch to turn on or off your DROID's Wi-Fi radio.

Adding a Hidden Network

If the network you want to connect to is not listed on the screen, it may be purposely hidden. If it is hidden it does not broadcast its name which is also known as its SSID. You need to touch Add a Network, type in the SSID, and choose the type of security that the network uses. You need to ask the network administrator for this information ahead of time.

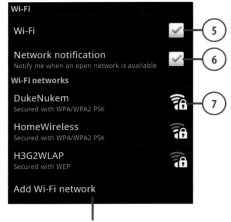

Touch to add a hidden network

6. Touch Network Notification to enable or disable the notification that tells you a new open Wi-Fi network is available.

7. Touch a Wi-Fi network to connect to it.

8. If the Wi-Fi network is secure, you receive a prompt to enter a password or encryption key.

9. Touch Connect to continue.

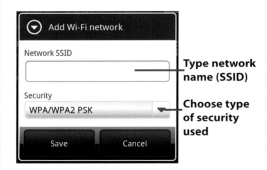

Type network name (SSID)

Choose type of security used

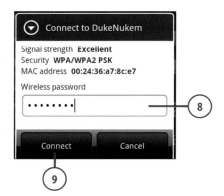

10. If all goes well you see the Wi-Fi network in the list with the word Connected under it.

Can't Connect to Wi-Fi?

If all does not go well, you might be typing the password or encryption key incorrectly. Verify this with the person who owns the Wi-Fi network. Sometimes there is a lot of radio interference that causes problems. Ask the person who owns the Wi-Fi network to change the channel it operates on and try again.

Indicates Wi-Fi is connected and shows signal strength

Wi-Fi Network Options

1. Touch a Wi-Fi network to reveal a pop-up that shows information about your connection to that network.

2. Touch Forget to tell your DROID to not connect to this network in the future.

3. Touch and hold on a Wi-Fi network to reveal two actions.

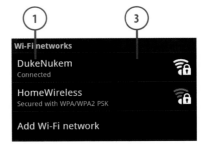

4. Touch to forget the Wi-Fi network and no longer connect to it.

5. Touch to change the Wi-Fi network password or encryption key that your DROID uses to connect to the network.

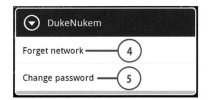

Advanced Wi-Fi Options

Your DROID allows you to configure a few advanced Wi-Fi settings that can actually help preserve your battery life.

1. Press the Menu button while on the Wi-Fi Settings screen and touch Advanced.

2. Touch to change the Wi-Fi sleep policy. This enables you to choose if your DROID automatically turns off Wi-Fi. Because Wi-Fi is more efficient that 3G, and is free, you should leave this set to Never.

3. Touch to enable or disable a Wi-Fi feature that boosts the Wi-Fi range of your DROID. This is useful if you are far away from the Wi-Fi router, but it does drain the battery faster so be careful.

4. Use this Wi-Fi MAC address if you need to provide ta network administrator with your MAC address in order to be able to use a Wi-Fi network.

5. Touch Use Static IP to force your DROID to use a static IP address and static network access settings.

6. Scroll down for more options.

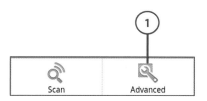

7. Touch to enable or disable a feature that letst your DROID choose its own IP address if one is not automatically assigned to it.

DLNA IP settings

DLNA auto-IP
Check to use DLNA when no DHCP server is available

⑦

Wi-Fi Is More Efficient and Free

Believe it or not, Wi-Fi is more efficient than a 3G cellular network. The more time you can keep the Wi-Fi radio on, and connected to a Wi-Fi network, the longer your battery lasts. If you have a valid data connection via Wi-Fi, your DROID actually stops using the cellular 3G network. So, while at home, why not set the Wi-Fi sleep policy to Never. Unless you have an unlimited data plan, it is also a good idea to use Wi-Fi whenever possible because it is free.

>>> Go Further

WHY USE A STATIC IP OR MAC?

Firstly what are an IP address and a MAC address? A MAC address is a number burned into your DROID that identifies its Wi-Fi adapter. This is called the physical layer because it is a physical adapter. An IP address is a secondary way to identify your DROID. Unlike a Physical Layer address or MAC address, the IP address can be changed anytime. Modern networks use the IP address when they need to deliver some data to you. Typically when you connect to a network, a device on the network assigns you a new IP address. On home networks, this is typically your Wi-Fi router.

In rare circumstances, the Wi-Fi network you connect to might not assign your DROID an IP address. In these circumstances, you will need to ask the network administrator to give you a static IP address.

Some network administrators use a security feature to limit who can connect to their Wi-Fi network. They set up their network to only allow connections from Wi-Fi devices with specific MAC addresses. If you are trying to connect to such a network, you will have to give the network administrator your MAC address, and he will add it to the allowed list.

Cellular Networks (GSM—DROID Pro and Incredible 2)

Your Verizon DROID Pro and Incredible 2 can connect to non-U.S. 3G (UMTS/HSDPA) cellular networks as well as U.S. CDMA networks. These are considered "Global" phones by Verizon, however they will not connect to AT&T or T-Mobile-US; they only connect to GSM networks outside the U.S.

Changing Mobile Settings

Your DROID has a few options when it comes to how it connects to cellular (or mobile) networks.

1. Press the Menu button and touch Settings.

2. Touch Wireless & Networks.

3. Touch Mobile Networks.

4. Touch Data Roaming to set whether your DROID connects to data while you are roaming outside your home carrier's network.

5. Touch to enable or disable a feature that plays a sound when you connect to a mobile network that is not your home network.

DROID Incredible 2

What Is an APN?

APN stands for Access Point Name. You normally don't have to make changes to APNs but sometimes you need to enter them manually to access certain features. For example, if you need to use tethering, which is where you connect your laptop to your DROID and your DROID provides Internet connectivity for your DROID, you might be asked by your carrier to use a specific APN. Think of an APN as a gateway to a service.

6. Touch to choose your preferred network (or Network Mode). Your choices are Global Mode, which means that your DROID uses CDMA while in the U.S. and GSM/UMTS while outside the U.S.; GSM/UMTS Mode, which forces your DROID to only use GSM/UMTS networks; and CDMA Mode, which forces your DROID to only use CDMA networks.

7. Touch to enable or disable always-on mobile data.

Can I Disable Always-on Mobile Data?

If you disable always-on mobile data, you can save on battery life, however you effectively kill the functionality of any app that needs to be connected all the time, such as Instant Messaging apps (Yahoo!, Google Talk, and so on), or apps like Skype. You also stop receiving email in real-time. When this feature is disabled, about five minutes after your DROID goes to sleep, it disconnects from the mobile data network, but it remains connected to the mobile voice network.

8. Touch to see specific GSM/UMTS options.

9. Touch to choose or add new APNs.

10. Touch to search for and select mobile operators, or set your DROID to choose them automatically.

DROID Pro

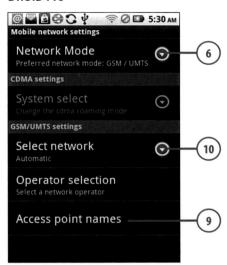

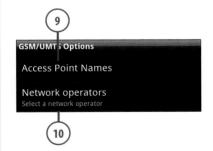

Cellular Networks (CDMA—All DROIDs)

Your Verizon DROID can connect to 2G (CDMA 1X), 3G (EVDO Rev A), and, in the case of the DROID CHARGE, 4G LTE cellular networks.

Changing Mobile Settings

Your DROID has a few options when it comes to how it connects to cellular (or mobile) networks.

1. Press the Menu button and touch Settings.

2. Touch Wireless & Networks.

3. Touch Mobile Networks.

4. Touch to enable or disable mobile data (DROID CHARGE only).

5. Touch Data Roaming to set whether your DROID connects to data while you are roaming outside your home carrier's network.

6. Touch to enable or disable a feature that plays a sound when you connect to a mobile network that is not your home network.

7. Touch to choose your preferred network (or Network Mode). Your choices are Global Mode, which means that your DROID uses CDMA while in the U.S. and GSM/UMTS while outside the U.S.; GSM/UMTS Mode, which forces your DROID to only use GSM/UMTS networks; and CDMA Mode, which forces your DROID to only use CDMA networks.

8. Touch to enable or disable always-on mobile data.

DROID 3, Pro, X2, Incredible 2

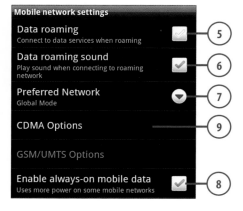

Can I Disable Always-on Mobile Data?

If you disable always-on mobile data, you can save on battery life, however you effectively kill the functionality of any app that needs to be connected all the time, like Instant Messaging apps (Yahoo, Google Talk, etc.), or apps like Skype. You will also stop receiving email in real-time. When this feature is disabled, about 5 minutes after your DROID goes to sleep, it will disconnect from the mobile data network, however it will remain connected to the mobile voice network.

9. Touch to select how your DROID roams. Your choices are Automatic and Home only. On the DROID CHARGE your choices are LTE Automatic and CDMA Mode.

DROID CHARGE

Virtual Private Networks (VPN)

Your DROID can connect to virtual private networks (VPNs), which are normally used by companies to provide a secure connection to their inside networks or intranets.

Adding a VPN

Before you add a VPN, you must first have all the information needed to set it up on your DROID. Speak to your network administrator and get this information ahead of time to save frustration. This information includes the type of VPN protocol used, type of encryption used, and the name of the host to which you are connecting.

1. Press the Menu button and touch Settings.

2. Touch Wireless & Networks.

3. Touch VPN Settings.

4. Touch to add a new PPTP, L2TP, or IPSec VPN.

5. Touch to add an Advanced IPSec VPN.

6. Touch the VPN technology your company uses.

DROID 3, CHARGE, Incredible 2

DROID X2, Pro

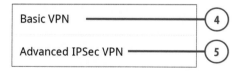

7. Go through each section and enter the information your network administrator gave you.

8. Press the Menu button and touch Save to save the VPN settings.

Connecting to a VPN

After you have created one or more VPN connections, you can connect to them when the need arises.

1. Follow steps 1–3 in the "Adding a VPN" task to navigate to the VPN Settings screen.

2. Touch the VPN you want to connect to.

3. Enter your username and password.

4. Touch Connect. After you're connected to the VPN, you can use your DROID's web browser and other applications normally, but you now have access to resources at the other end of the VPN tunnel, such as company web servers or even your company email.

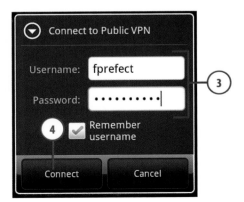

>>> Go Further

EDIT OR DELETE A VPN

You can edit an existing VPN or delete it by touching and holding on the name of the VPN. A window pops up with a list of options.

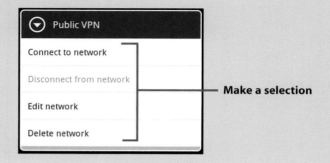

Make a selection

3G Mobile Hotspot

Your DROID has an application called 3G Mobile Hotspot or Mobile Hotspot that enables you to use Wi-Fi to share your DROID's 3G Internet connection with other devices. The exact number of devices depends on the DROID you have. This feature only works with an optional extra monthly fee, which covers the extra data you use when this feature is enabled. Because each of the DROIDs does this differently, each device has a task of its own.

3G Mobile Hotspot (DROID Incredible 2)

1. Touch to launch 3G Mobile Hotspot.

2. Enter a unique name for the Wi-Fi network you create when you enable this feature.

3. Touch to select the type of security you want to use on the Wi-Fi network, or select no security. It is a good idea to use security so that the people who connect to your hotspot have encrypted data going over the air.

4. Enter the password or key you want to use on your hotspot to encrypt the data.

5. Touch to start your mobile hotspot.

After Your Hotspot Is Started

After you start your hotspot, provide the router name (SSID), security, and password to anyone who you want to allow to connect. In the notification bar you see the 3G Mobile Hotspot icon indicating that your hotspot is working. Touch Manage Users to see who is connected.

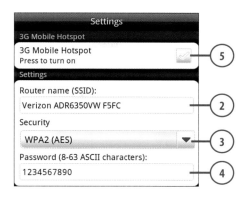

Indicates your hotspot is working

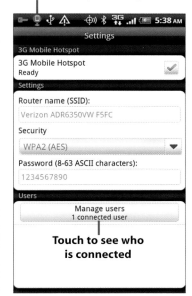

Touch to see who is connected

Advanced Settings

Sometimes you want to have some more control over your 3G Mobile Hotspot, such as changing the Wi-Fi channel you use or changing how IP works on your hotspot.

1. Touch to launch 3G Mobile Hotspot.

2. Touch to choose the Wi-Fi channel that your hotspot uses or leave it set to Auto.

3. Touch to change the number of minutes after which your hotspot automatically disables itself if it detects no activity.

4. Touch to change your hotspot's LAN settings. Only people with an understanding of IP should change the LAN settings.

5. Touch to choose the IP address of your Wi-Fi hotspot.

6. Touch to change the subnet mask of the IP address.

7. Touch to enable or disable the ability for your Wi-Fi hotspot to automatically assign IP address to devices that connect.

8. Touch to choose the starting IP address that your Wi-Fi hotspot uses when it automatically assigns addresses.

9. Touch to reset the LAN settings.

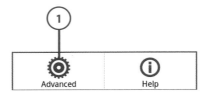

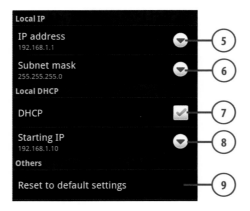

Mobile Hotspot (DROID X2)

1. Touch to launch Mobile Hotspot.

2. Touch to set the maximum number of devices that can connect to your Hotspot and also limit who can connect based on their MAC address.

3. Touch to change the number of minutes after which your hotspot automatically disables itself if it detects no activity.

4. Touch to start your Wi-Fi hotspot.

5. Touch to configure your Wi-Fi hotspot after it starts.

6. Enter a unique name for the Wi-Fi network you create when you enable this feature.

7. Touch to select the type of security you want to use on the Wi-Fi network, or select no security. It is a good idea to use security so that the people who connect to your hotspot have encrypted data going over the air.

8. Enter the password or key you want to use on your hotspot to encrypt the data.

9. Touch to change the Wi-Fi channel used by your Wi-Fi hotspot.

10. Touch to save the changes.

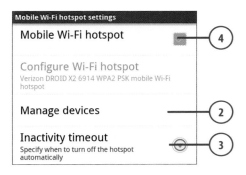

After Your Hotspot Is Started and Configured

After you start your Hotspot and configure it, provide the router name (SSID), security, and password to anyone who you want to allow to connect. In the notification bar you see the Mobile Hotspot icon indicating that your hotspot is working. Touch to configure the Mobile Hotspot settings.

Indicates your hotspot is working ———

Touch to see who is connected

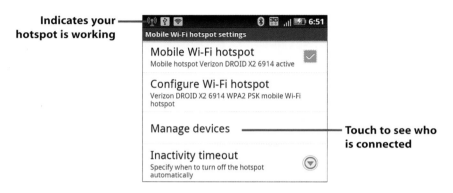

Mobile Hotspot (DROID CHARGE)

1. Touch to launch Mobile Hotspot.

2. Touch to start your Mobile Hotspot. The first time you use the Mobile Hotspot feature, you are asked to configure it.

3. Touch Configure.

4. Touch Yes.

5. Touch Automatic. Your DROID automatically sets up your hotspot.

6. Touch Turn Wi-Fi Off.

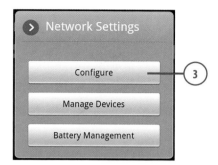

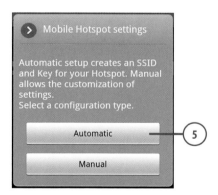

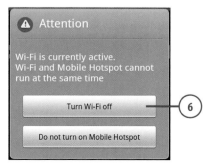

7. Touch Turn on Mobile Hotspot.

After Your Hotspot Is Started

After you start your hotspot, provide the router name (SSID), security, and password to anyone who you want to allow to connect. In the notification bar you see the Mobile Hotspot icon indicating that your Hotspot is working.

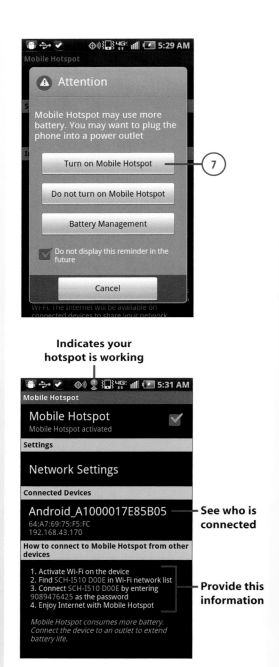

Advanced Settings

Sometimes you want to have some more control over your 3G Mobile Hotspot such as changing the network name and the type of security used.

1. Touch Network Settings.

2. Touch to change the number of minutes after which your hotspot automatically disables itself if it detects no activity.

3. Touch to see who is connected to your hotspot and also limit who can connect based on their MAC address.

4. Touch to change your hotspot's network name and security type.

5. Touch Yes, Switch to Manual.

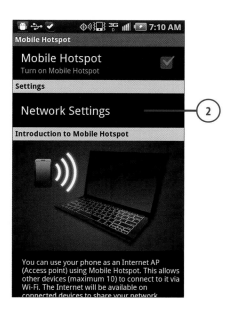

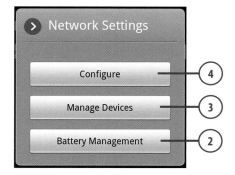

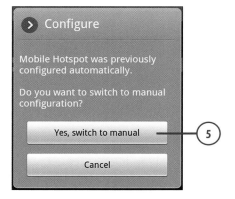

6. Enter a new network name or SSID.

7. Touch Next.

8. Touch to change the type of security used by your hotspot.

9. Touch to change the security password.

10. Touch Done.

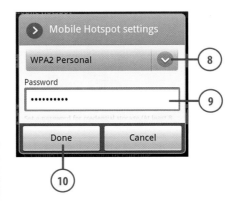

3G Mobile Hotspot (DROID Pro)

1. Touch to launch 3G Mobile Hotspot.

2. Press the Menu button and touch Advanced to set up your hotspot before starting it.

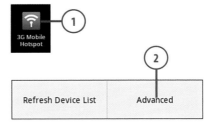

3. Touch Wi-Fi AP mode.

4. Touch to choose the type of security used by your hotspot.

5. Enter a security password.

6. Touch to change the Wi-Fi channel used by your hotspot.

7. Scroll down for more settings.

8. Touch to change the number of minutes of inactivity after which your DROID waits before automatically disabling the hotspot.

9. Touch Save.

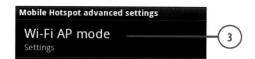

Mobile Hotspot advanced settings

Wi-Fi AP mode — 3
Settings

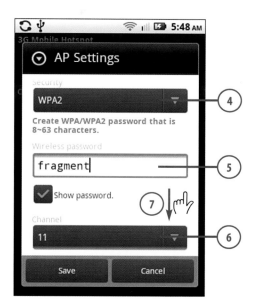

10. Press the Back button to return the main Hotspot screen.

11. Touch to enable your Hotspot.

After Your Hotspot Is Started

After you start your hotspot, provide the router name (SSID), security, and password to anyone who you want to allow to connect. In the notification bar you see the 3G Mobile Hotspot icon indicating that your hotspot is working.

Hotspot Sucks Power

When you use 3G Mobile Hotspot, it is a good idea to keep your DROID plugged in to the wall or to a car charger. This is because 3G Mobile Hotspot uses a lot of power and CPU cycles and drains your battery very quickly.

**Indicates your
hotspot is
working**

See who is connected

In this chapter, you learn about your DROID's email applications for Gmail and other email accounts such as POP3, IMAP, and even Microsoft Exchange. Topics include the following:

→ Adding a Gmail account
→ Adding a Microsoft Exchange account
→ Sending and receiving email
→ Working with attachments

5

Emailing

Your DROID Incredible 2 has two email programs: the Gmail application, which only works with Gmail, and the Mail application that works with POP3, IMAP, and Microsoft Exchange accounts. Your DROID 3, Pro, and X2 have three email programs: the Gmail application, which only works with Gmail, the Email application, which only works with POP3 and IMAP, and the Messaging application that provides a universal Inbox for all email except Gmail. Your DROID CHARGE has a Gmail application for Gmail only and an Email application for everything but text messages.

Gmail

When you first set up your DROID, you set up a Gmail account. The Gmail application enables you to have multiple Gmail accounts, which is useful if you have a business account and a personal account.

Adding a Gmail Account

When you first set up your DROID, you added your first Gmail account. The following steps describe how to add a second account.

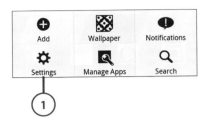

1. Press the Menu button and touch Settings.

2. Touch Accounts. On some DROIDs this is called Accounts & Sync.

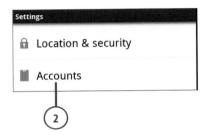

3. Touch Add Account.

4. Touch Google.

5. Touch Next.

6. Touch Sign in.

What If I Don't Have a Second Google Account?

If you don't already have a second Google account, but want to set one up, in step 6, touch Create. Your DROID walks you through the steps of choosing a new Google account.

7. Enter your existing Google account name. This is your Gmail address.

8. Enter your existing Google password.

9. Touch Sign in.

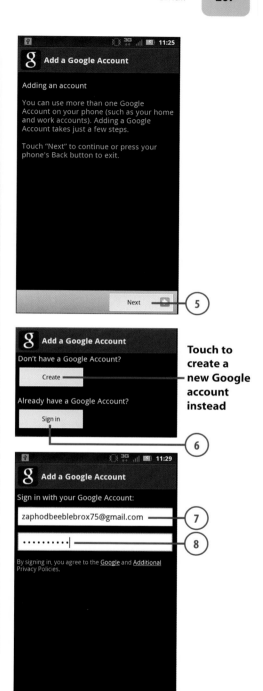

Touch to create a new Google account instead

10. Select what components of your Google account you want to synchronize with your DROID.

What Are All These Other Google Components?

The list of Google components you see in step 10 depends on the version of the Google client that is preinstalled on your DROID. The DROID 3 has the most extensive list including the ability to sync Google+ notifications and Google Books. Most DROIDs only show Gmail, Google Calendar, and Google Contacts.

11. Touch Finish.

12. Touch Finish Setup.

Why Multiple Google Accounts?

You are probably wondering why you would want multiple Google accounts. Isn't one good enough? Actually it is not that uncommon to have multiple Google accounts. It can be a way to compartmentalize your life between work and play. You might run a small business using the one account, but email only friends with another. Your DROID supports multiple accounts but still enables you to interact with them in one place.

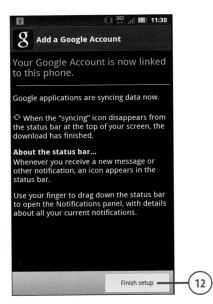

Composing Gmail Email

Now that you have at least one Gmail account set up, you can start sending and receiving email using one of the Gmail accounts.

Touch to look at a different label

1. Touch the Gmail icon on the Home screen.

2. Press the Menu button and touch Compose.

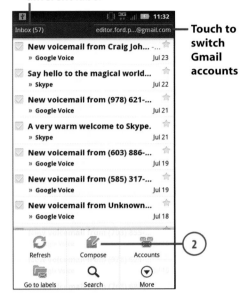

Touch to switch Gmail accounts

Stars and Labels

Gmail allows you to use stars and labels to help organize your email. In most email clients you can create folders in your Inbox to help you organize your emails. For example you might create a folder called "emails from the boss" and move any emails you receive to that folder. Gmail doesn't use the term folders; it uses the term labels instead. You can create labels in Gmail and choose an email to label. When you do this, it actually moves it to a folder with that label, but to you, the email has a label distinguishing it from other emails. And email that you mark with a star is actually just getting a label called "starred." But when viewing your Gmail, you see the yellow star next to an email. People normally add a star to an email as a reminder of something important.

3. Touch to change the Gmail account from which the message is being sent.

4. Type names in the To field. If the name matches someone in your Contacts, the name is displayed and you can touch it to select it.

5. Type a subject.

6. Enter the body of the email.

7. Press the Menu button and touch Attach to attach one or more files to the email.

8. Select the location of the files you want to attach. The Gallery contains all pictures and video whereas the Files option enables you to browse anywhere on your DROID.

9. Touch to remove an attachment.

10. Touch to send the email.

Touch to save as a draft

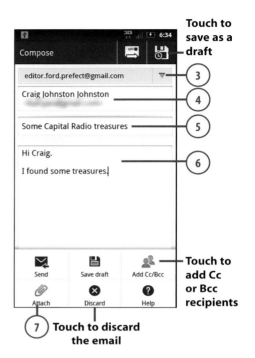

Touch to add Cc or Bcc recipients

Touch to discard the email

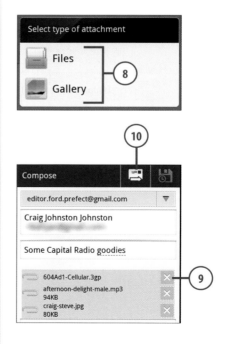

Reading Gmail Email

It seems obvious enough, but this section covers how to handle attachments and a few other tricks.

1. Touch an email to open it.

2. Touch to add the email to the Gmail star label.

3. Touch to reply to the sender of the email.

4. Touch to show more response options, including reply all and forward.

5. Touch to archive the email.

6. Touch to delete the email.

7. Touch to see the previous email in your inbox.

8. Touch to see the next email in your inbox.

9. Scroll down to read the rest of the email and see any attachments.

Indicates attachments

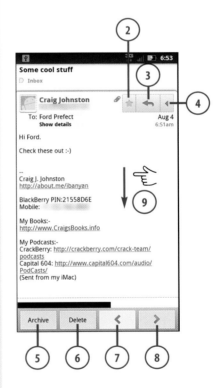

10. Touch to download the attachment.

11. Touch to preview the attachment.

Only Download Pictures

For some reason, the Gmail app only
allows you to download and save
picture attachments on your DROID.
You can preview other attachment
types but you can't download them.

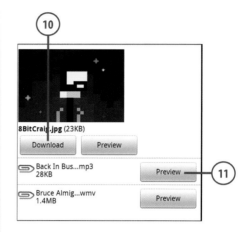

12. Press the Menu button to see
more options.

13. Touch Mark Unread to return the
email to unread status.

14. Touch Mute to block emails from
this sender.

15. Touch to add the email to the
Gmail star label.

16. Touch to report this email as
spam.

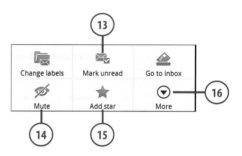

What Happens to Your Spam?

When you mark an email in Gmail as Spam, two things happen. Firstly it gets a
label called Spam. Secondly a copy of that email is sent to Gmail's Spam servers
so they are now aware of a possible new Spam email that is circulating around
the Internet. Based on what the servers see for all Gmail users, they block that
Spam email from reaching other Gmail users. So the bottom line is, always mark
Spam emails as Spam because it helps all of us.

Gmail Settings

You can customize the way Gmail accounts work on your DROID by changing the email signature and choosing which labels synchronize.

1. Touch to choose the Gmail account for which you want to change the settings.

2. Press the Menu button and touch More.

Email Signature

An email signature is a bit of text that is automatically added to the bottom of any emails you send from your DROID. It is added when you compose a new email, reply to an email, or forward an email. A typical use for a signature is to automatically add your name and maybe some contact information at the end of your emails. Email signatures are sometimes referred to as email footers.

3. Touch Settings.

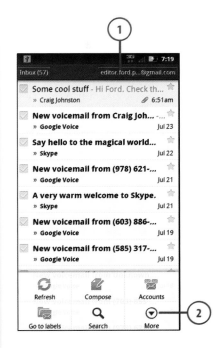

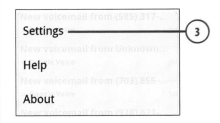

4. Touch Signature to type a signature that will appear at the bottom of emails you send from this Gmail account.

5. Touch to select which actions you perform on emails prompt you for confirmation.

6. Touch to make Reply All the default action when replying to emails. Normally only Reply is used. Reply All replies to the sender and all recipients.

7. Touch Auto-advance to select which screen your DROID must show after you delete or archive and email. Your choices are Newer Conversation, Older Conversation, and Conversation List.

8. Touch to select the size of the text used when reading emails. Your choices range from Tiny to Huge.

9. Touch to enable or disable batch operations. When this is enabled, there are always check boxes next to emails in the conversation list. This enables you to select more than one email and take action on it.

10. Touch to clear the email search history.

11. Touch to select which of your Gmail labels are synchronized to your DROID.

12. Scroll down for more settings.

What Are Conversations?

Conversations are Gmail's version of email threads. When you look at the main view of the Gmail app, you are seeing a list of email conversations. The conversation might have only one email in it, but to Gmail that's a conversation. As you and others reply to that original email, Gmail groups those emails in a thread, or conversation.

13. Touch to enable or disable notifications for this Gmail account. Notifications are displayed in the status bar.

14. Touch to select the ringtone that plays when new email arrives for this Gmail account.

15. Touch to enable or disable vibration when new email arrives for this account.

16. Touch to enable or disable the notify once feature. This makes your DROID notify you once per batch of new incoming emails instead of once per email.

Setting Up Email Accounts

Before you start working with the Messaging (Universal Inbox) or Email applications, you must first add email accounts.

Adding a New Corporate Email Account (DROID 3, X2, Pro, CHARGE)

If you use your DROID as your only phone, you might want to have both personal and corporate email on it. Your DROID uses the ActiveSync protocol for corporate accounts. This is built into Microsoft Exchange but can be available through mail gateways such as Lotus Traveler and Novell DataSync.

1. Press the Menu button and touch Settings.

2. Touch Accounts (Accounts & Sync on the DROID CHARGE).

DROID 3, Pro, X2

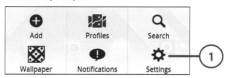

DROID CHARGE

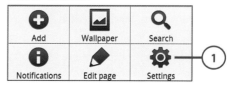

DROID 3, Pro, X2

DROID CHARGE

3. Touch Add Account.

4. Touch Corporate Sync (Corporate on the DROID CHARGE).

DROID 3, Pro, X2

DROID CHARGE

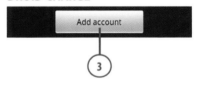

DROID 3, Pro, X2

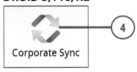

DROID CHARGE

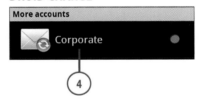

5. Enter your corporate email address.

6. Enter your corporate network password. This is the password you use to log in to your company's network every day.

7. Enter your company's Active Directory Domain name.

8. Enter your company network login username.

9. Touch Next.

DROID 3, Pro, X2

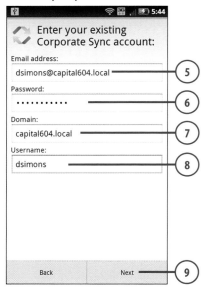

DROID CHARGE

Touch to enter more information

10. Touch Done (DROID 3, Pro, X2 only). Steps 11 to 20 are for the DROID CHARGE only.

Password copied
DROID CHARGE from previous screen

7 — Domain — capital604.local

8 — Username — dsimons

Enter your company's ActiveSync server name — Password ·········· — Exchange server capital604.local

Keep this checked — ☑ Use secure connection (SSL)

Only check this if told to — ☑ Accept all SSL certificates

9 — Next

DROID 3, Pro, X2

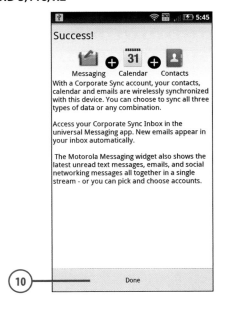

Success!

Messaging Calendar Contacts

With a Corporate Sync account, your contacts, calendar and emails are wirelessly synchronized with this device. You can choose to sync all three types of data or any combination.

Access your Corporate Sync Inbox in the universal Messaging app. New emails appear in your inbox automatically.

The Motorola Messaging widget also shows the latest unread text messages, emails, and social networking messages all together in a single stream - or you can pick and choose accounts.

10 — Done

11. Touch to choose how email is delivered during peak times. Your choices are to have email pushed to your DROID in real time, poll for mail at certain intervals, or set it to manual so that email is only retrieved when you run the Email app.

12. Touch to choose how email is delivered during off-peak times. Your choices are to have email pushed to your DROID in real time, poll for mail at certain intervals, or set it to manual so that email is only retrieved when you run the Email app.

13. Touch to enable or disable notifications when email arrives for this account.

14. Touch to enable or disable synchronizing email with this account.

15. Touch to enable or disable synchronizing contacts with this account.

16. Touch to enable or disable synchronizing the calendar with this account.

17. Touch Next.

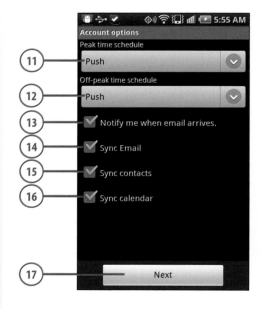

18. Give the account a name.

19. Touch Done.

20. Touch Activate to allow your company's email administrator control over your DROID. This allows your company to impose restrictions on your DROID such as enforcing a password, but it also allows the administrator to remotely wipe it when you leave the company.

DROID CHARGE Only

DROID CHARGE Only

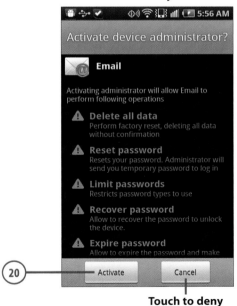

Touch to deny administrator control

Corporate Account Settings (DROID 3, X2, Pro)

After you set up a corporate account, you can change the way it functions on your DROID.

1. Press the Menu button and touch Settings.

2. Touch Accounts.

3. Touch the Corporate account you set up in the previous section.

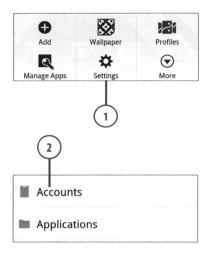

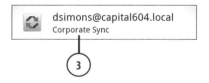

4. Change your Corporate domain and username if you need to.

5. Change your corporate network password if you need to.

6. Touch to enable or disable synchronizing your corporate calendar with your DROID.

7. Touch to enable or disable synchronizing your corporate contacts with your DROID.

8. Touch to enable or disable synchronizing your corporate email with your DROID.

9. Touch to change your corporate email address.

10. Touch to change your ActiveSync server if you need to.

11. Change the name of your corporate account. For example you can call it "Work mail."

12. Touch to enable or disable connecting to your corporate mail using encryption. You should leave this enabled at all times.

13. Scroll down for more settings.

14. Touch to enable or disable certificate verification. You should always leave this enabled unless your administrator tells you otherwise.

15. Touch to select how far back in time emails must be synchronized to your DROID.

16. Touch to remove your corporate account.

17. Touch to save the settings.

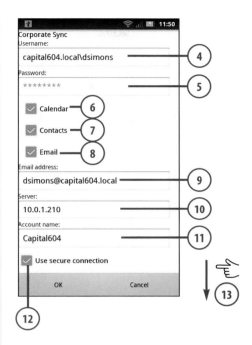

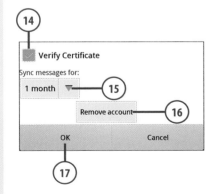

18. Press the Back button to return to the main Settings screen.

19. Touch Battery & Data Manager.

20. Touch Data Delivery.

21. Touch to enable or disable synchronizing data in the background. You should leave this on all the time otherwise your email will never be delivered in real-time.

22. Touch to enable or disable connecting to data services when roaming off your home network.

23. Touch to enable or disable connecting to any data network. If you disable this, no app that uses data, including email will function.

24. Touch Email and Corporate Sync to see more settings.

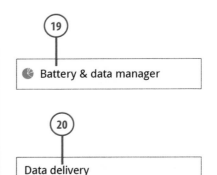

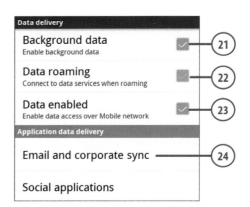

25. Touch to change how you are notified when you receive a new email. You can change the ringtone used, choose whether to also vibrate the DROID, or disable notifications.

26. Touch to select how your corporate email is delivered. You can make it synchronize only when you are on a Wi-Fi network, send it in real-time (push), put it on a polling interval, select how attachments are downloaded, and choose which folders to synchronize.

27. Touch to select the text size used when you read emails.

28. Touch to select how many lines of the email you see when looking at the list of emails, and choose whether to always have multi-select on, which enables you to work with more than one email at a time.

29. Touch to change the look of your text when you compose an email and change your email signature.

30. Touch to enable or disable name suggestions. This enables your DROID to suggests names as you type them in to the To field.

31. Touch to change the default email account to use when you compose new email.

32. Touch to set your out of office notification schedule and message.

33. Scroll down to see more settings.

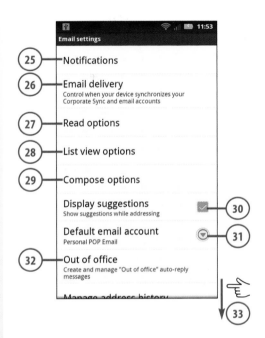

34. Touch to manage your email address history. This allows you to delete names and addresses from being suggested as you type names into the To field.

35. Touch to enable or disable Smart Forwarding. If your company uses Exchange 2010, this feature is available. It saves on wireless data by keeping the attachments from an email you forward on the mail server instead of sending them via your DROID.

36. Press the Back button to save your settings.

Corporate Account Settings (DROID CHARGE)

After you set up a corporate account, you can change the way it functions on your DROID CHARGE.

1. Press the Menu button and touch Settings.

2. Touch Accounts & Sync.

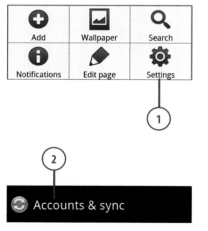

3. Touch the Corporate account you set up in the earlier section.

4. Touch Accounts Settings.

5. Touch to change the account name, for example you could call it "work email" or office email."

6. Touch to enable or disable this as the default email account to use when sending emails from your DROID.

7. Touch to choose whether you want to always Cc or Bcc yourself when you send emails from your DROID.

8. Touch to choose how far back in time to synchronize email to your DROID.

9. Touch to empty the trash folder in your mailbox. Some companies impose a mailbox size limit and when you reach it you can no longer send emails. Emptying your Trash folder can clear some space.

10. Touch to manually configure when peak and off-peak time. You can also choose how email is delivered to your DROID in peak and off-peak times.

11. Touch to set up your out-of-office period and the message to send while you are out. When you set your Out of Office settings, they are changed on your mailbox so they are in effect no matter if your DROID is on or off.

12. Scroll down for more settings.

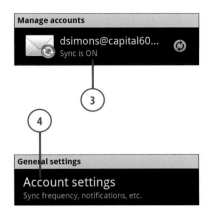

13. Touch to choose the size of emails and attachments to retrieve from the server. You can set this to any size, or you can choose All, which retrieves the whole email and all attachments no matter how large they are.

14. Touch to view the ActiveSync security policies and restrictions set by your email administrator. You cannot change them, only view them.

15. Touch to enable or disable corporate email notifications.

16. Touch to select the ringtone to use when you are notified of a new corporate email.

17. Touch to also vibrate when playing the new email notification.

18. Scroll down for more settings.

19. Touch to change corporate mail account settings such as password, ActiveSync server name, and so on.

20. Touch to enable or disable synchronizing your corporate email to your DROID.

21. Touch to enable or disable synchronizing your corporate contacts to your DROID.

22. Touch to enable or disable synchronizing your corporate calendar to your DROID.

23. Touch to add a signature to all outgoing emails sent from your DROID.

24. Touch to change the signature.

25. Touch to enable or disable a confirmation message to pop up when you delete email.

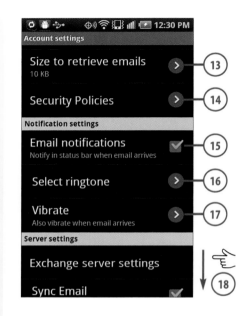

Adding a New POP3 or IMAP Account (DROID 3, Pro, and X2)

It's time to add a POP3 or IMAP account. These are typically used by hosted email systems such as Yahoo! or Hotmail.

1. Press the Menu key and touch Settings.

2. Touch Accounts.

3. Touch Add Account.

4. Touch Email.

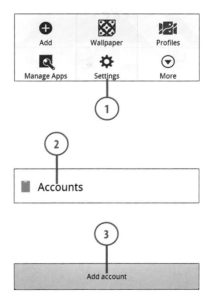

5. Enter your email address.

6. Enter your password.

7. Touch Next.

8. If all goes well, skip to step 18.

9. Touch Set up Manually if you are prompted.

Why Manual?

Your DROID tries to figure out the settings to set up your email account. This works most of the time when you are using common email providers such as Yahoo! or Hotmail and others. It also works with large ISPs such as Comcast, Road Runner, Optimum Online, and so on. It might not work for smaller ISPs, or in smaller countries, or if you have created your own website and set up your own email. In these cases, you need to set up your email manually.

10. Touch Incoming Server.

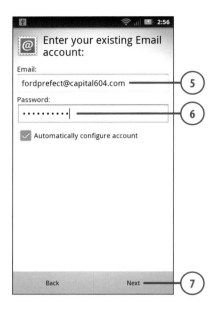

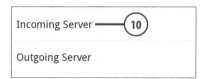

11. Touch to choose between POP3 and IMAP accounts.

12. Enter your account's POP3 or IMAP server name.

13. Touch OK.

Where Can I Find This Information?

If you need to manually set up your email account, you must have a few pieces of information. Always check your ISP's, or email service provider's, website, and look for instructions on how to set up your email on a computer or smartphone. This is normally under the support section of the website.

14. Touch Outgoing Server.

15. Enter your account's outgoing mail server name.

16. Touch OK.

Touch for advanced settings such as port numbers

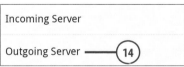

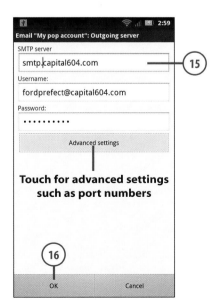

Touch for advanced settings such as port numbers

17. Touch OK.

Username and Password

On the Incoming Server and Outgoing Server screens, your user-name and password should already be filled out because you typed them in earlier. If not, enter them.

18. Touch Done.

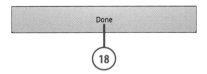

Be Secure If You Can

If your mail provider supports email security such as SSL or TLS, you should strongly consider using it. If you don't, emails you send and receive go over the Internet in plain readable text. Using SSL or TLS encrypts the emails as they travel across the Internet so nobody can read them. Set this under the Advanced settings for the Incoming and Outgoing Servers.

Adding a New POP3 or IMAP Account (DROID CHARGE)

Unlike the other DROIDs, on the DROID CHARGE you can only set up a POP3 or IMAP account from within the Email app.

1. Touch the Email icon in the Launcher.

2. Press the Menu button and touch Accounts after the Email app launches.

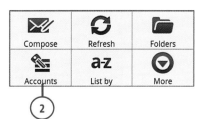

3. Touch Add Account.

4. Touch Others.

5. Enter your email address.

6. Enter your password.

7. Touch Next.

8. If all goes well, skip to step to step 22, otherwise continue manually setting up the account.

9. Select whether the account is a POP3 or IMAP account.

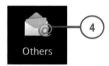

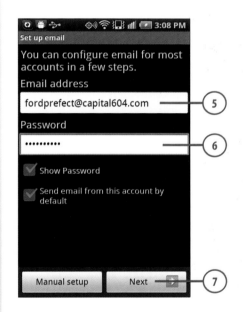

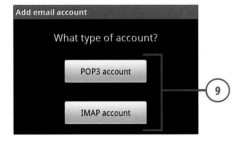

10. Enter your account's username.

11. Enter your account's password.

12. Enter the incoming POP3 or IMAP server name.

13. Change the incoming server port number if necessary.

14. Select a type of encryption to use if your account supports it.

15. Enter the outgoing server name.

16. Scroll down for more settings.

17. Change the outgoing server port number if necessary.

18. Touch to use an encryption method if your account supports it.

19. Touch to enable or disable the setting that sends your account login information to the outgoing server.

20. Touch to enable or disable using the Verizon email gateway. When enabled, it allows Verizon to push emails to you in real time.

21. Touch Next.

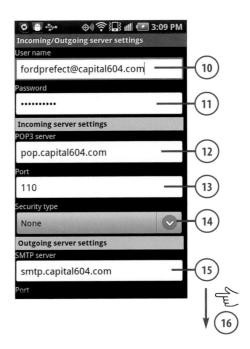

Verizon Gateway

Normally, POP3 accounts do not push email to your DROID in real time. They rely on your DROID polling for new email at certain intervals. Verizon offers a service that, when selected, polls your email account very frequently and delivers email to your DROID in near real time. This saves on your DROID's battery life because it's not doing the actual polling.

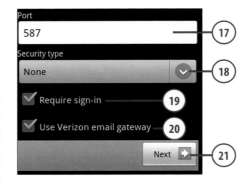

22. Touch to select whether your email is pushed to your DROID in real-time, or retrieved on a regular interval.

23. Touch to enable or disable a notification when new email arrives for this account.

24. Touch Next.

25. Enter a name for your account—something such as "personal email."

26. Enter your full name as you would like it displayed.

27. Touch Done.

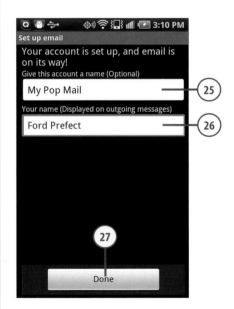

Adding a New Microsoft Exchange Account (DROID Incredible 2)

If you use your DROID as your only phone, you might want to have both personal and corporate email on it. The DROID Incredible 2 uses an application called Mail for all non-Gmail email.

1. Press the Menu button and then touch Settings.

2. Touch Accounts & Sync.

3. Touch Add Account.

4. Touch Exchange ActiveSync.

5. Enter your corporate email address.

6. Enter your corporate network password.

7. Touch Next.

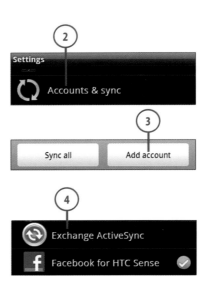

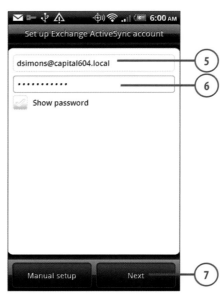

8. Enter your company's ActiveSync server. You can try using webmail.<your company's domain>.<domain extension> or auto discover.<your company's domain>.<domain extension> and if that doesn't work, ask your administrator.

9. Enter your company's Active Directory domain name.

10. Enter your corporate network login username.

11. Check to use a secure encrypted connection between your DROID and your company's mail server. This is highly recommended.

12. Touch Next.

13. Select what to synchronize to your DROID.

14. Touch have your corporate email pushed to your DROID in real time.

15. Touch to have your DROID poll for email every 15 minutes.

16. Touch Next.

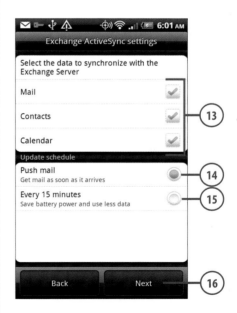

17. Enter a name for this account. You can use something like "work mail."

18. Touch Finish Setup.

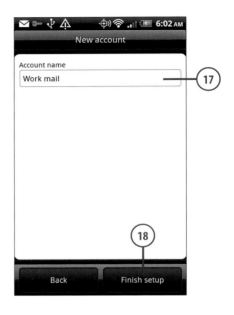

Exchange Account Settings (DROID Incredible 2)

After you set up an Exchange ActiveSync account, you can change the way it functions on your DROID. For example you might want it to synchronize more email.

1. Press the Menu button and touch Settings.

2. Touch Accounts & Sync.

3. Touch Exchange ActiveSync.

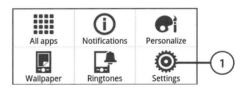

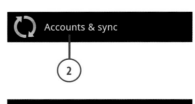

4. Select what to synchronize to your DROID.

5. Touch to enable or disable updating your corporate data when you open the Mail or Calendar apps.

6. Touch to change when mail is delivered, during peak times and off-peak times, and if it is sent in real time (push) or on a polling interval.

7. Touch to synchronize your corporate data.

8. Touch to set more ActiveSync settings.

9. Touch to edit your ActiveSync connection settings like your username or password.

10. Touch to enable or disable synchronize your corporate data while roaming outside your home carrier network.

11. Touch to choose how to resolve any conflicts between your DROID and your mailbox. You can replace the item on your DROID with the one from your mail server, or keep both items even if one of them is incorrect. An example of a conflict is an email that has changed on your DROID but not in your mailbox.

12. Touch to choose how far back in time to synchronize your corporate email.

13. Touch to set the size limit of emails downloaded to your DROID.

14. Touch to change the format of emails created on your DROID.

15. Touch choose whether to include attachments when synchronizing email to your DROID. You can choose not to include attachments, include attachments smaller than a set size, or include all attachments no matter how large they are.

16. Touch to set how far back in time your corporate calendar synchronizes to your DROID.

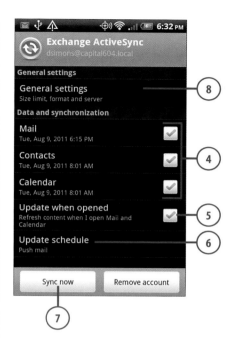

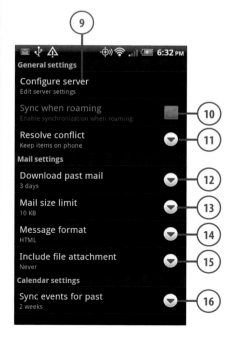

Adding a New POP3 or IMAP Account (DROID Incredible 2)

You can also add a POP3 or IMAP account to your DROID Incredible 2. These accounts are typically used by hosted email systems such as Yahoo! or Hotmail.

1. Press the Menu button and touch Settings.

2. Touch Accounts & Sync.

3. Touch Add Account.

4. Touch Mail.

5. Enter your email address.

6. Enter your password.

7. Touch Next.

8. If all goes well you should skip to step 19; otherwise continue to step 9 to enter some information about your account manually.

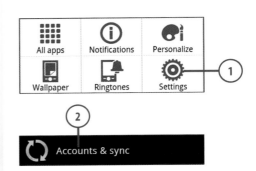

Why Manual?

Your DROID tries to figure out the settings to set up your email account. This works most of the time when you are using common email providers such as Yahoo! or Hotmail. It also works with large ISPs such as Comcast, Road Runner, Optimum Online, and so on. It might not work for smaller ISPs, in smaller countries, or if you have created your own website and set up your own email. In these cases, you need to set up your email manually.

9. Choose the type of email account this is—POP or IMAP.

10. Enter the account's username.

11. Enter the incoming (POP or IMAP) server name.

Be Secure If You Can

If your mail provider supports email security such as SSL or TLS, you should strongly consider using it. If you don't, emails you send and receive go over the Internet in plain readable text. Using SSL or TLS encrypts the emails as they travel so nobody can read them.

12. Touch to choose a type of encryption to use when synchronizing mail with this account.

13. Enter a different incoming mail server port if necessary.

Where Can I Find This Information?

If you need to manually set up your email account, you must have a few pieces of information. Always check your ISP's, or email service provider's, website, and look for instructions on how to set up your email on a computer or smartphone. This is normally under the support section of the website.

14. Touch Next.

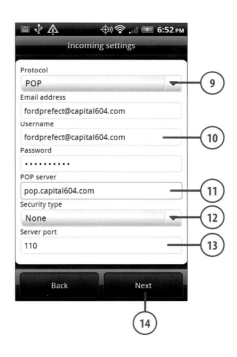

15. Enter your outgoing (SMTP) server name.

16. Touch to choose a type of encryption to use when sending mail with this account.

17. Change the outgoing server port if necessary.

18. Touch Next.

19. Enter a name for this account.

20. Enter your name as you would like it displayed when sending mail from this account.

21. Touch Finish Setup.

Working with the Messaging Application (DROID 3, Pro, X2)

Now that you have added two new accounts, you can start using the Messaging application. Everything you do in the email application is the same for every email account. The Messaging application also supports text messaging.

Using Combined or Separate Inboxes

The Messaging application enables you to work with your different email accounts separately or in a combined Inbox (Universal Inbox).

1. Touch the Messaging icon in the Launcher.

2. Touch to compose a new message.

3. Touch to open only your text messages.

4. Touch to open only your corporate email.

5. Touch to open only your Facebook email.

6. Touch to open only your POP3 or IMAP email.

7. Touch to open the Universal Inbox that has all email and text messages in a single view.

8. Touch or press the Menu button to reveal the Menu.

9. Touch Edit Universal Inbox to change which accounts are included in the Universal Inbox view.

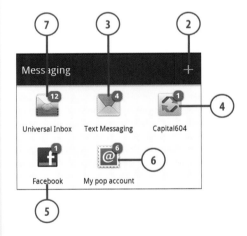

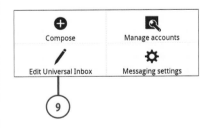

Working in the Universal Inbox

The Universal Inbox is the easiest way to work with all your email and text message accounts because they are all in one place.

1. Touch Universal Inbox.

2. Press the Menu button to reveal more options.

3. Touch to refresh the Universal Inbox view.

4. Touch a checkbox to select a message. After you have one or more messages selected you can take action on them such as deleting them or, if they are all from the same account, moving them to a folder.

5. Touch select all messages.

6. Touch to compose a new message.

7. Indicates the message is from your corporate email account.

8. Indicates the message is from your Facebook account.

9. Indicates the message is a text message.

10. Indicates the message is from your personal POP3 or IMAP email account.

11. Touch a message to read it.

12. Touch to change your account settings.

Universal Inbox

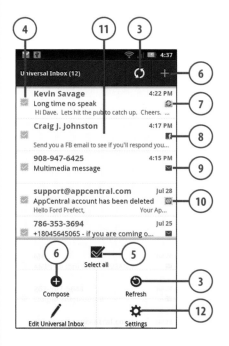

Composing Email

No matter if you are looking at the Universal Inbox or a single account's inbox, composing email is the same.

1. Touch to compose a new message.

2. Select the account from which you want to send the message.

3. Enter one or more recipients. As you type, your DROID tries to guess who you want to address the message to. If you see the correct name, touch it to select it. This includes names stored on your DROID and in your company's corporate address book.

4. Enter a subject.

5. Type the message.

6. Change the formatting of text by selecting the text and touching one of the formatting icons.

7. Touch to start a bulleted list.

8. Touch to insert an emoticon (smiley face).

9. Touch to change the font being used, font size, and font's foreground and background color. To change font properties, first select the text you want to change.

10. Touch to send the message.

Adding Attachments

Before you send your message you might want to add one or more attachments.

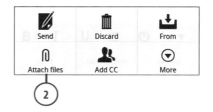

1. Press the Menu button.
2. Touch Attach Files.
3. Choose the type of file to attach.
4. Touch Send.

Mark a Message As Urgent

To mark a message as urgent, before you send it, press the Menu button, touch More, and then touch Importance. This enables you to mark a message as High, Normal, or Low Importance.

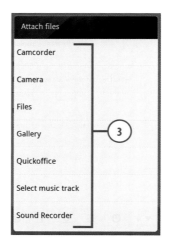

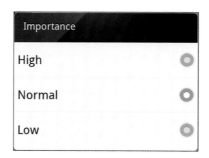

Reading Email

Reading messages in the Email application is the same regardless of which account the email has come to.

1. You receive a notification in the status bar letting you know that there is a new email.

2. Pull down the status bar.

3. Touch the notification to be taken directly to that new email.

4. Touch to see all the email recipients.

5. Touch to reply, reply-all, or forward the email.

6. Touch to delete the email.

7. Touch to preview and download the attachments.

8. Press the Menu button to see more options.

9. Touch to move the email to a folder.

10. Touch to flag the message.

11. Touch to mark the message as unread.

12. Touch to save a copy of the email on your DROID.

New email notification

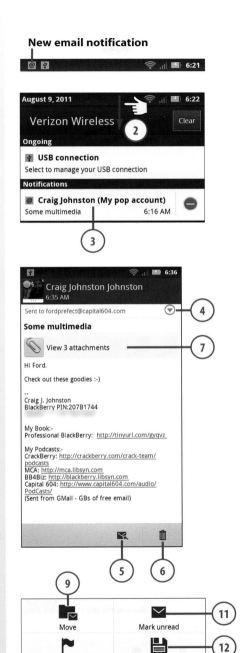

Previewing and Downloading Attachments

While you read an email, you can preview any attachments and download them to your DROID.

1. Touch to see the list of attachments.

2. Touch to preview the attachment.

3. Touch to save the attachment to your DROID.

4. Press the Menu button and touch Save all to save all attachments to your DROID.

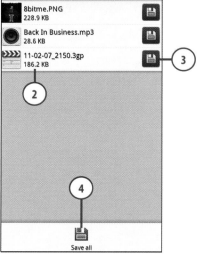

Working with the Email Application (DROID CHARGE)

Now that you have added two new accounts, you can start using the Email application. Everything you do in the Email application is the same for every email account. Unlike the DROID 3, Pro, and X2, the DROID CHARGE uses an application called Email for all non-Gmail email accounts.

Navigating the Email Application

Before you learn how to compose or read emails, you should become familiar with the Email application.

1. Touch to launch the Email app icon.

2. Touch to switch between email accounts. The more accounts you have added to your DROID CHARGE, the more tabs there are.

3. Touch to change the folder you are viewing.

4. Touch an email to open it.

5. Touch to mark the email as flagged.

6. Check boxes next to emails to select more than one. Then you can take actions against multiple emails at once, such as Mark as Read, Flag, Delete, or Move to a new folder.

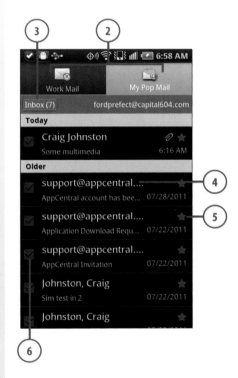

7. Press the Menu button to see more options.

8. Touch to compose a new email.

9. Touch to refresh the mailbox view and manually load any new emails.

10. Touch to switch to a different folder in the mailbox you are viewing.

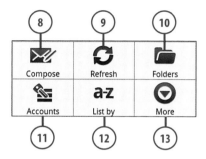

11. Touch to work with and add accounts supported by the Email app.

12. Touch to change the sort order of the emails listed in the mailbox view. You can choose to sort by date, sender, read/unread (where read and unread emails are grouped together), and title of email.

13. Touch to see more options.

14. Touch to change the account settings for the mailbox you are viewing. If you are viewing your corporate mailbox, you can empty the Trash folder and set your Out Of Office. Read more about account settings earlier in this chapter.

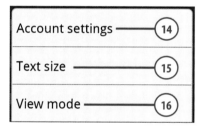

15. Touch to change the size of the text being used to display messages.

16. Touch to change the view mode. You can switch between the Standard View and the Conversation View. The Conversation View groups emails by thread or conversation.

Composing Email

1. Press the Menu button and touch Compose.

2. Enter one or more recipients. As you type, your DROID tries to guess who you want to address the message to. If you see the correct name, touch it to select it. This includes names stored on your DROID and in your company's corporate address book

3. Touch to change which account to use to send the email.

4. Enter a subject.

5. Type the message.

6. Touch to cancel the message.

7. Touch to attach files to the message.

8. Press the Menu button for more options.

9. Touch to add Carbon Copy (Cc) and Blind Carbon Copy (Bcc) recipients.

10. Touch to spell check your email.

11. Touch to save the email as a draft so you can come back to it later and finish it.

12. Touch to add information from contacts you have stored on your DROID CHARGE or add birthday information from stored contacts to the email.

13. Touch to set the email as low, normal, or high priority.

14. Touch to show the text formatting tool. This only works if you are using the Swype keyboard.

15. Press the Back button to close the menu.

16. Touch to send the message.

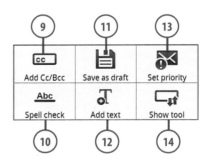

Adding Birthday and Other Info to Emails

On your DROID CHARGE you can take advantage of a feature that enables you to insert information from contacts you have stored on your DROID into an email. To do this, before you send an email, press the Menu button and touch Add Text. Choose Namecard to insert multiple bits of information for a name-card, or Calendar to insert only someone's birthday information. Choose the namecard to use, select the information to insert, and touch Submit. The information from the namecard you selected will be inserted into the email.

Formatting Email Text

If you use the Swype keyboard you can add formatting to your emails. To do this, press the Menu button and touch Show Tool. After the formatting tool bar appears, select words and use the tools to change the formatting to bold, italic, underline, and even choose between default and monospace fonts. You can also insert bulleted lists.

Change text formatting

Insert bulleted list

Select text

Add Attachments

Before you send your message you might want to add one or more attachments.

1. Press the Menu button and then touch Attach.

2. Choose the type of file to attach.

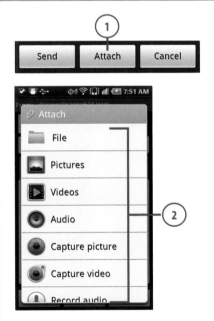

List of attachments

3. Touch to remove an attachment.

4. Touch Send.

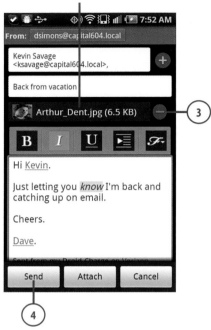

Reading Email

Reading messages in the Email application is the same regardless of which account the email has come to.

1. You receive a notification in the status bar letting you know that there is a new email.

2. Pull down the status bar.

3. Touch the notification to go directly to that new email.

New email notification

4. Touch to see all the email recipients.

5. Touch to reply to the email.

6. Touch to delete the email.

7. Touch to preview and download the attachments.

8. Press the Menu button to see more options.

9. Touch to move the email to a folder.

10. Touch to add the sender to your contacts.

11. Touch to mark the message as unread.

12. Touch to forward the email.

13. Touch to see more actions.

14. Touch change the background of the body of the email to beige or leave it as white.

15. Touch to change the size of the font used when you're reading emails.

16. Touch to create a new calendar entry and insert the email into the body of that calendar entry.

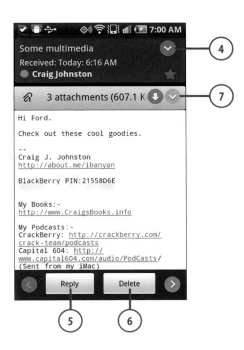

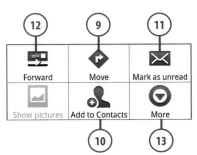

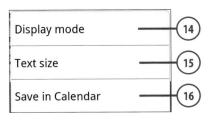

Previewing and Downloading Attachments

While you read an email, you can preview any attachments and download them to your DROID.

1. Touch to see the list of attachments.

2. Touch to preview the attachment.

3. Touch to save the attachment to your DROID.

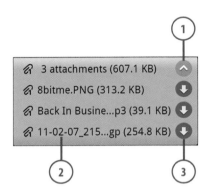

Working with the Email Application (DROID Incredible 2)

Now that we have added two new accounts, we can start using the Mail application. Everything you do in the Mail application is the same for every email account.

Navigating the Main Screen

The Mail application has multiple views and can switch between email accounts so you can keep your work and home email separated.

1. Touch to launch the Mail app.

2. Touch to choose which email accounts to display in the Inbox view. You can see all accounts, or you can choose to see just one account at a time.

3. Indicates how many unread emails there are in the current view.

4. Touch to compose a new email.

5. Touch to open an email.

6. Indicates the email is flagged for follow-up.

7. Indicates an email in your POP3 or IMAP account.

8. Indicates an email in your corporate mail account.

9. Select multiple emails to take actions on them like mark unread or delete.

10. Switch to the conversations view where emails are grouped by their email threads or conversations.

11. Touch to see only emails that are marked as favorites.

12. Touch to see only unread emails.

13. Scroll right to see more mailbox views.

14. Touch to see only emails that are flagged for follow-up.

15. Touch to see only calendar invites.

16. Press the Menu button to see more options.

17. Touch to refresh the mailbox view and load any new emails.

18. Touch to change the sort order of the list of emails. You can choose between many different sort orders including sorting by date, priority, subject, sender, and size.

19. Touch to switch to a different mailbox folder.

20. Touch to see all emails or only emails that have been marked with a star.

21. Touch to see more options.

22. Touch create a new calendar invitation. This opens the calendar application, which is covered in Chapter 8, "Date, Time, and Calendar."

23. Touch to add a new email account to your DROID.

24. Touch to set your corporate out of office message for the dates when you'll be away.

Indicates an urgent email

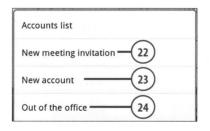

Composing Email

No matter which mail account you
are working with, composing email is
the same.

1. Touch Compose Mail.

2. Type in the recipients. If the
   DROID recognizes names you
   type as being in your Contacts,
   you see a list of names to choose
   from. Commas need to separate
   recipients' addresses, but luckily
   the application adds them for
   you. If you type in an address
   manually, when you hit space, a
   comma is added. If you select an
   address, a comma is added.

3. Touch to search for recipients on
   your corporate network.

4. Enter a subject.

5. Type your email.

6. Touch to discard the email.

7. Touch to save the email as a draft
   and finish it later.

8. Touch to send the email.

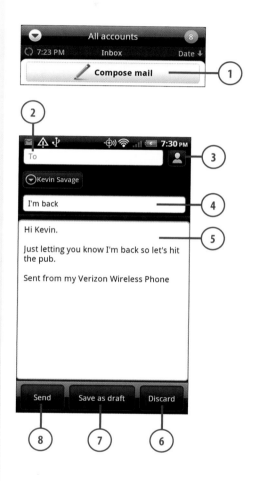

It's Not All Good

Corporate Name Search Is Kludgy

The DROID Incredible 2 has a kludgy way of searching for corporate contacts. Unlike other DROIDs that search your corporate directory as you type someone's name into the To field, these DROIDs make you perform a multi-step process. To search your corporate directory, touch the icon as shown in step 3. Type in part or all of the person's name. Touch the big search icon. Once you see your search results, select the correct name and touch Done.

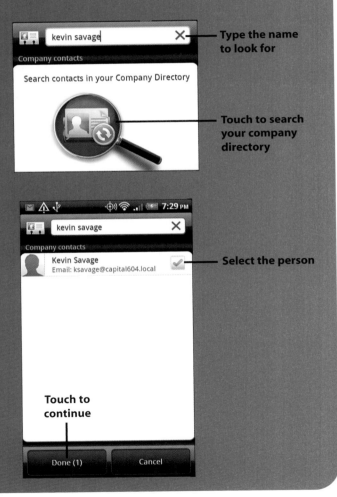

Type the name to look for

Touch to search your company directory

Select the person

Touch to continue

8. Press the Menu button to see more options.

9. Touch to set the priority of the email.

10. Touch to show the Cc and Bcc fields in your email and add recipients to them if needed.

11. Touch to send your email.

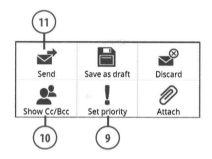

Adding Attachments

Before you send your message you might want to add one or more attachments.

1. Press the Menu button and touch Attach.

2. Select the type of file to attach. Some of your choices are to attach contact cards (vCards), your location (your current GPS coordinates), pictures, videos, and audio.

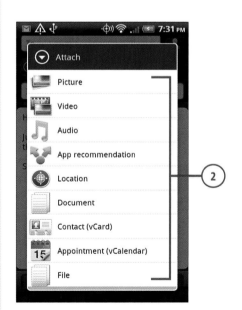

3. Touch Send.

**Touch to hide or show
the list of attachments**

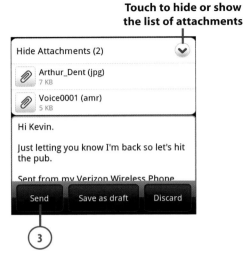

Reading Email

Reading messages in the Mail application is the same regardless of
which account the email has come
to.

New email notification

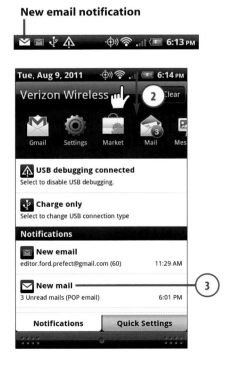

1. When a new email arrives you see
   a notification in the status bar.

2. Pull down the status bar.

3. Touch the email notification.

4.   Touch to reveal all recipients.

5.   Touch to reveal any attachments.

6.   Touch to reply to the sender and any other recipients.

7.   Touch to reply only to the sender.

8.   Touch to move to the previous email (newer).

9.   Touch to move to the next email (older).

10.   Press the Menu button to reveal more options.

11.   Touch to return to the received messages view.

12.   Touch to the conversations view, highlighting this message in that view.

13.   Touch to compose a new message.

14.   Touch to mark the message you are reading as unread.

15.   Touch to delete the message you are reading.

16.   Touch to see more actions.

17.   Touch to forward the message.

18.   Touch to move the message to a folder.

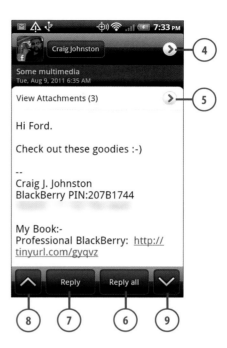

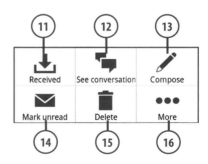

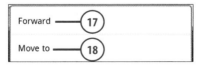

In this chapter, you learn about browsing the World Wide Web and using the browser capabilities of your DROID. Topics include the following:

→ Bookmarking websites
→ Sharing websites with your friends
→ Keeping track of sites you have visited
→ Using GPS and browsing together

Browsing the Web

Your DROID has a fully featured web browser for a smartphone. In fact the experience using the DROID's browser is similar to using a desktop browser, just with a smaller screen. You can bookmark sites, hold your DROID sideways to fit more onto the screen, and even share your GPS location with sites.

Navigating with the Browser

Let's dive right in and cover how to run the browser and use all of its features. Your DROID's browser can be customized, shares your GPS location, enables you to bookmark sites, and keeps your browsing history.

1. Touch Browser on the Home screen. On the DROID Incredible 2, this icon is called Internet.

2. Touch to type in a new web address. Some websites move the web page up to hide the address field. When this happens, you can drag the web page down to reveal the address bar again.

3. Touch to see the Browser bookmarks and add a bookmark. (See the next section for information on setting bookmarks.)

4. Press the New Window button to see more options.

DROID Incredible 2 Bookmarks Icon

On the DROID Incredible 2 this icon is not there. To reach the bookmarks on this DROID, touch the Menu button and then touch Bookmarks.

5. Touch to see a list of open browser windows (see more about windows later in this chapter).

6. Touch to refresh the current web page.

7. Touch to go forward one page. This option is only available if you used the Back button to go back one page.

8. Touch to see more options.

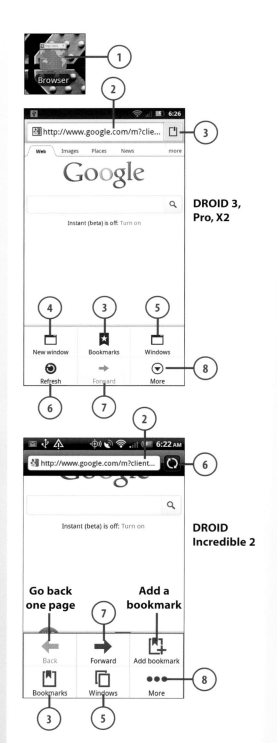

DROID 3, Pro, X2

DROID Incredible 2

9. Touch to find text on the current web page.

10. Touch to select text on the current web page.

11. Touch to see the information about the current web page.

12. Touch to share the link to the current web page with friends via email, text message, and other methods.

13. Touch to see a list of any files you have downloaded from web pages.

14. Touch to see your browser history.

15. Touch to change the browser settings.

16. Touch to subscribe to a RSS feed. This option is greyed out when the web site you are viewing has no RSS feeds (DROID CHARGE only).

17. Touch to add the current web page to your Home screen as a shortcut (DROID CHARGE only).

18. Touch to add the current page as a bookmark (DROID CHARGE only).

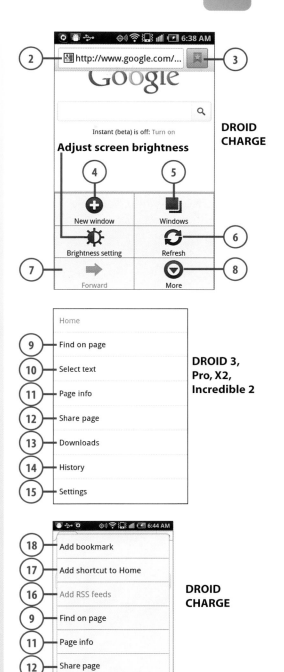

Browser Tricks

Your DROID has some unique tricks to help you browse regular websites on a small screen.

1. Rotate your DROID on its side, also called landscape orientation. Your DROID automatically switches the screen to landscape mode.

Zoomed out —————

Zoomed in —————

2. Double-tap the screen to zoom in and out.

Pinch to Zoom

An alternative way to zoom, which enables you to actually zoom in much further, is to place your thumb and forefinger on the screen and spread them apart to zoom in, and move them back together to zoom out.

Managing Bookmarks, Most Visited, and History

Your DROID enables you to bookmark your favorite websites, but it also keeps track of where you have browsed and can show you your most visited sites, or let you see what website you visited three months ago.

DROID 3, Pro, X2

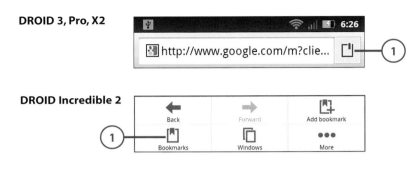

DROID Incredible 2

DROID CHARGE

1. Touch the bookmarks icon to see your bookmarks, most visited sites, and browser history.

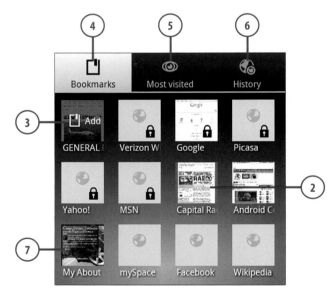

2. Touch a bookmark to load that website.
3. Touch to add the website you are currently viewing as a new bookmark (DROID 3, Pro, X2, and CHARGE only).

DROID X2 Alternative View

On the DROID X2, if you rotate it onto its side (landscape mode), the view switches to a "cover flow" view, which enables you to swipe left and right to move through the bookmarks, most visited sites, and browser history.

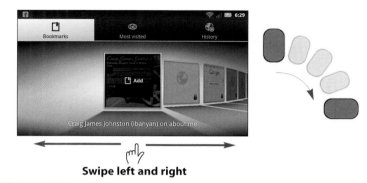

Swipe left and right

4. Touch to view the browser bookmarks.

5. Touch to see your most visited websites.

6. Touch to see your browser history.

7. Touch and hold a bookmark to see the bookmark actions.

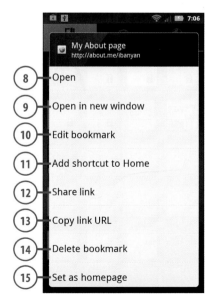

8. Touch to open the bookmarked website.

9. Touch to open the bookmarked website in a new browser window.

10. Touch to edit the bookmark.

11. Touch to add the bookmark to your DROID's Home screen as a shortcut.

12. Touch to share the bookmarked website's link with friends using email, text messaging, and other methods.

13. Touch to copy the bookmarked website's link to your DROID's clipboard. After it is in the clipboard, you can paste it into any other screen, such as the body of an email message.

14. Touch to delete the bookmark.

15. Touch to make this bookmark your Browser's Home page. The Home page is the page that loads when you first open the Browser.

Adding a Bookmark (DROID 3, Pro, X2, CHARGE)

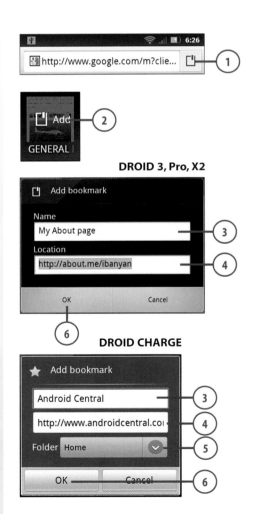

1. Touch the Bookmarks icon while viewing the web page you want to bookmark.

2. Touch the Add icon on the top-left of the Bookmarks screen.

3. Change the bookmark name if you want to. It defaults to the web page's title.

4. Edit the web page link if you want to or leave it as it (normally best).

5. Select a folder to save the book-mark to (DROID CHARGE only).

6. Touch OK.

DROID CHARGE Bookmark Folders

Your DROID CHARGE has a unique feature that enables you to organize your bookmarks into folders. In step 5, when you create a bookmark, you can select which folder to put it into. To create a new bookmark folder, press the Menu button and touch Create folder. The folder displays in the bookmark thumbnails. To move an existing bookmark into a folder you must edit the bookmark and change the folder it is saved to.

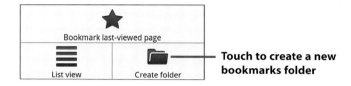

Touch to create a new bookmarks folder

Adding a Bookmark (DROID Incredible 2)

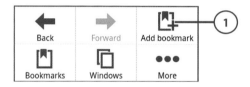

1. Press the Menu button and touch Add Bookmark while viewing the web page you want to bookmark.

2. Change the bookmark name if you want to. It defaults to the web page's title.

3. The URL will show up in the Location box.

4. Touch to add tags to the bookmark. If you don't want to add tags, skip to step 7.

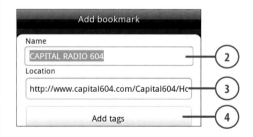

5. Select the tags you want to associate with this bookmark.

6. Touch Done.

7. Touch Done.

Type new tags and touch the + icon

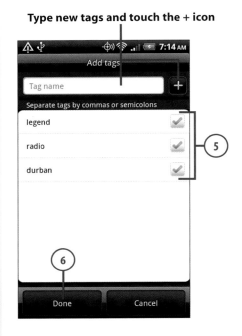

Associated tags

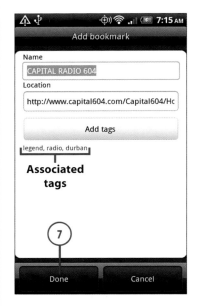

DROID Incredible 2 Tags

Your DROID Incredible 2 has a unique feature that enables you to associate tags with your bookmarks. By associating tags, which are like search terms, with your bookmarks, you can use the Tags view to see bookmarks organized by tags.

Touch a tag to see bookmarks that have that tag

Touch to see the Tags view

Most Visited

Most viewed websites are assigned their rankings based on how often you view each one over time.

DROID 3, Pro, X2, CHARGE

1. Touch Most Visited. This tab is called Most Viewed on the DROID CHARGE. On the DROID Incredible 2 you touch the heart icon to see Most viewed.

2. Touch a website name to load it.

DROID Incredible 2

History

The History screen retains all websites that you have visited and groups them by date. This is useful if you want to find a website you know you visited on a particular date, but can't remember the exact address.

1. Touch History. On the DROID Incredible 2, the icon is at the bottom of the screen.

2. Touch a date to expand it and see the websites you visited on that day or during that month.

3. Touch a website to visit the website.

Clear History

To clear your browser history, press the Menu button and then touch Clear History to clear all traces of your web browsing history.

DROID 3, Pro, X2, CHARGE

DROID Incredible 2

Indicates the site is bookmarked

Subscribing to News Feeds (DROID 3, Pro, X2)

As you browse different websites, you might notice that some of them cause a symbol to be shown on the address bar. That symbol means a website has a Real Simple Syndication (RSS) feed. You can subscribe to those feeds and receive news updates without having to visit the websites themselves.

1. Touch the RSS icon.

2. Select either the News app or the News widget to handle the RSS feed.

3. Touch Subscribe.

Subscribing to News Feeds (DROID CHARGE)

When you subscribe to RSS feeds on your DROID CHARGE, it subscribes to them in the Google Reader service, which is one of your Google account services.

1. Touch the RSS icon.

2. Touch the RSS feed you want to subscribe to. Some web sites have multiple feeds, one for posts and one for comments as an example.

3. Enter your Google password.

4. Touch Sign In.

5. Touch Subscribe.

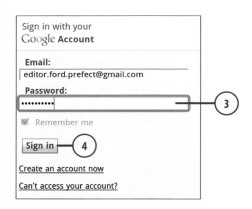

Subscribing to News Feeds (DROID Incredible 2)

When you subscribe to RSS feeds on your DROID Incredible 2, it subscribes to them in the Google Reader service which is one of your Google account services.

1. Touch the RSS icon.

2. Touch the RSS feed you want to subscribe to. Some websites have multiple feeds—for example, one for top stories and one for recent stories.

3. Touch Add.

Where Can You See the News Feeds?

If you have a DROID 3, Pro, X2, or Incredible 2, you must use the News app. It is preinstalled on your DROID. If you use a DROID CHARGE, you must log in to the Google website and go to the Reader tab. Alternatively, you can search for a Google RSS Reader app in the Android Market and install it. RSS Reader apps synchronize any RSS feeds to which you have subscribed in Google Reader to your DROID CHARGE.

Sharing Your GPS Location

Your DROID has a built-in GPS radio that enables it to tell exactly where it is on planet Earth. If you allow websites to see this information, they can become even more useful. For example, when you search Google your search results are filtered to include local stores, or a website can use your location to provide driving directions.

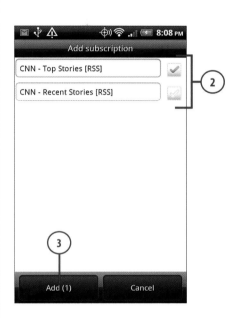

Always share your location with this website

Share your location with this website

Managing Multiple Windows

Your DROID can have multiple web pages open at one time, each in a differ-ent window. This enables you to work with multiple websites at once and switch between them. Here is how to open and work with multiple windows.

1. Press the Menu button and touch Windows.

2. Touch to add a new Browser win-dow.

DROID 3, Pro, X2

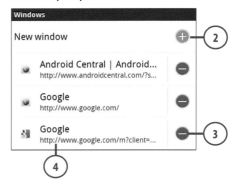

DROID CHARGE

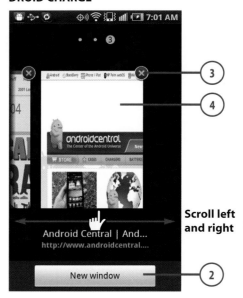

Scroll left and right

3. Touch to close a Browser window.

4. Touch a window name to switch to that window.

DROID Incredible 2

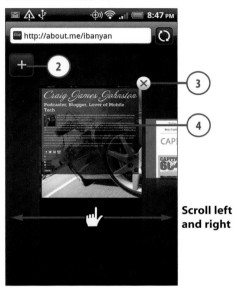

Scroll left and right

Customizing Browser Settings

Your DROID's browser is customizable. Here are the different settings you can adjust. The order of these settings depends on the model of DROID you are using.

1. Touch or press the Menu button and touch More.

2. Touch Settings.

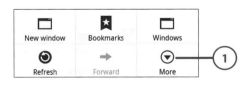

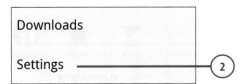

3. Touch Text Size to change the text size used when rendering web pages. Choices range from Tiny to Huge. The default is Normal.

4. Touch to change the default zoom level when opening web pages. Choices are Far, Medium, and Close. The default zoom is Medium.

5. Touch Open Pages in Overview to turn off page overview. Page overview is when a page appears zoomed out to an overview of the page, as opposed to the page being viewed at 100% zoom.

6. Touch Text Encoding to choose a different encoding option. Use this to select text encoding for Japanese and other characters.

7. Touch to block pop-up windows. Pop-up windows are almost always advertisements, so keeping this enabled is a good idea; however, some websites might require that you turn off pop-up blocking.

8. Touch Load Images to enable or disable image loading. When Load Images is disabled, the Browser loads web pages with no images, which makes the pages load faster.

9. Touch Auto-fit Pages to make web pages automatically fit the screen horizontally.

10. Touch Landscape-only Display to force all web pages to display in landscape mode.

11. Scroll down to see more settings.

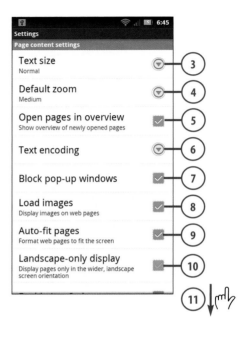

12. Touch to enable or disable JavaScript. JavaScript is used on many web pages for formatting and other functions, so you might want to leave this enabled.

13. Touch to enable or disable Browser plug-ins. Plug-ins enhance the standard functionality of the Browser.

14. Touch Open in Background to force loading new windows in the background. When this is enabled, a new window does not automatically become the current window; you have to switch to it to make it the current window.

15. Touch Set Home Page to manually set the Browser home page to the current web page, or type in any web page address.

16. Touch to clear everything from the Browser cache, history, cookies, form data, and passwords.

17. Touch to clear the Browser cache. The Browser cache is used to store data and images from websites you visit so that the next time you go back, some of the web page can be loaded straight out of memory.

18. Touch to clear your Browser history. This clears the history of websites you have visited on your DROID.

19. Touch Accept Cookies to enable or disable accepting cookies. Browser cookies are used by websites to personalize your visit by storing information specific to you in the cookies.

20. Scroll down for more settings.

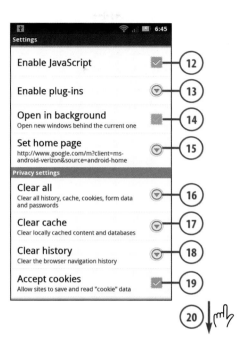

21. Touch to clear all cookie data.

22. Touch Remember Form Data to enable or disable remembering form data. Form data is information you enter into forms on web pages, such as your name, phone number, or email address. When you enable this feature you can automatically fill in fields on forms.

23. Touch to clear all previously entered web form data.

24. Touch Enable Location to enable or disable the ability for websites to access your GPS information.

25. Touch to clear location settings for all websites you have visited.

26. Touch Remember Passwords to enable or disable the Browser's ability to remember the usernames and passwords you enter on different websites.

27. Touch to clear any passwords you have previously entered while on websites.

28. Touch Show Security Warnings to enable or disable website security warnings. Your DROID can warn you if a website you are visiting appears to have a security violation of some kind.

29. Scroll down for more settings.

30. Touch to select the search engine used when you type in search terms. Your choices are Google, Yahoo, and Bing.

31. Touch Website Settings to see which websites you share your location information with, and disable this feature for certain websites.

32. Touch Reset to Default to return the Browser to the out-of-the-box state and clear all Browser data.

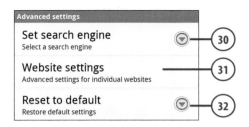

In this chapter, you learn how to work with text and multimedia messages. Topics include the following:

→ Creating text and multimedia messages
→ Attaching files to multimedia messages
→ Saving received multimedia attachments
→ Working with text messages on your SIM card

Text and Multimedia Messaging

Short Message Service (SMS), also known as text messages, have been around for a long time and are still used today as the primary form of communication for many phone users. Multimedia Message Service (MMS) is a newer form of text messaging that can contain pictures, audio, and video. Your DROID is at the top of its game when it comes to SMS and MMS.

Text Messaging Application

The Text Messaging application (called Messaging on the DROID CHARGE and Messages on the DROID Incredible 2) is what you use to work with SMS and MMS. This application has all the features you need to compose, send, receive, and manage these messages.

1. Touch the Text Messaging icon on the Home screen.

2. Touch to compose a new text message.

3. Touch to show the Quick Contact bar.

4. Touch a text message to open it.

5. Press the Menu button to see options.

6. Touch to select and delete multiple messages.

7. Touch to see the Messaging app settings.

8. Touch to search for a text message (DROID CHARGE only).

9. Touch to see draft text messages. These are messages that you didn't complete and saved as drafts (DROID Incredible 2 only).

10. Touch to compose a text message to groups of people. These are groups that you created in the People application. Please see Chapter 1, "Contacts," for more about groups (DROID Incredible 2 only).

DROID 3, Pro, X2   **DROID CHARGE**

DROID Incredible 2

Messages

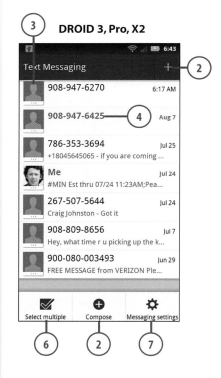

DROID CHARGE

DROID Incredible 2

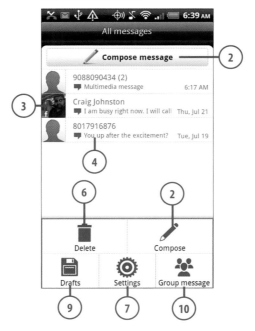

Text Messaging Application Settings (DROID 3, Pro, X2)

You manage how your SMS and MMS messages are handled through the settings of the Messaging and Messages applications.

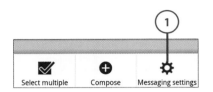

1. From the main screen, press the Menu button and touch Messaging Settings.

2. Touch to enable or disable delivery reports for SMS or text messages that you send.

Delivery Reports

When you enable delivery reports, your DROID keeps track of the text message and provides confirmation that it was successfully delivered to the recipient(s). If you touch and hold on the text message you are able to read the delivery report details.

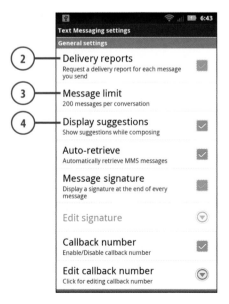

3. Touch to change the message limit, or the maximum number of text messages, per message thread. The maximum number you can type is 999. When the limit is reached, messages within the thread are deleted using the first in first out (FIFO) method.

4. Touch to enable or disable name suggestions when addressing a text message.

5. Touch to enable or disable auto-retrieval of your messages. You probably want to leave this enabled. If you disable this then your messages do not arrive in real time, which almost defeats the purpose of SMS and MMS.

6. Touch to enable or disable adding a signature to all text messages you send. Remember that a signature adds to the number of characters in a message.

7. Touch to type in your signature if you enabled it in step 6.

8. Touch to enable or disable adding your callback number to text messages you send.

9. Touch to change your callback number if you enabled sending it in step 8.

10. Scroll down to see more settings.

What Is a Callback Number?

If you choose to send your callback number along with your text messages then your DROID appends the number you want to use as a callback number to the end of the message. On some phones the recipient might see [CB] and your number at the end of messages. Most phones know how to handle the callback number and do not display it as part of the text message. However, not many phones honor the callback number. Many times even if you set the callback number option and type in the phone number you want to use, the recipient's phone simply calls your DROID's phone number. On your DROID you can actually change the callback number so that if a recipient tries to call you back, and their phone honors the callback number, you can direct them to a different phone number.

11. Touch to enable or disable receiving a notification that a new text message has arrived. The notification is displayed in the status bar.

Don't Auto Retrieve While Roaming

When you travel to other countries disable auto-retrieval of messages because auto-retrieving messages when you're roaming can result in a big bill from your provider. International carriers love to charge large sums of money for people traveling to their country and using their network. The only time it is a good idea to leave this enabled is if your carrier offers an international SMS or MMS bundle where you pay a flat rate upfront before leaving.

12. Touch to choose a different ringtone to be played when new text message arrives.

13. Touch to enable or disable vibration when new messages are received.

Messaging Application Settings (DROID CHARGE)

1. Press the Menu button and touch Settings.

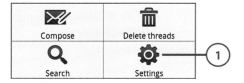

2. Touch to enable or disable the deletion of messages after the limits per thread are reached. This option works in conjunction with the options shown in steps 3 and 4.

3. Touch to change the message limit, or the maximum number of text messages, per message thread. The maximum number you can type is 999.

4. Touch to change the message limit, or the maximum number of multimedia messages, per message thread. The maximum number you can type is 999.

5. Touch to enable or disable delivery reports for SMS or text messages that you send.

6. Touch to manage messages stored on your SIM card. This is only valid for GSM DROID models or DROIDs that can roam onto GSM networks.

7. Touch to enable or disable delivery reports for multimedia messages (MMS).

8. Scroll down to see more options.

9. Touch to enable or disable auto-retrieval of your messages. You probably want to leave this enabled. If you disable this then your messages do not arrive in real time, which almost defeats the purpose of SMS and MMS.

10. Touch to enable or disable auto-retrieval of your messages when you are roaming on a carrier that is not your own. Because roaming fees can apply, you may choose to leave this disabled.

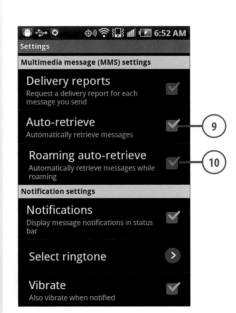

11. Touch to enable or disable whether a message is displayed in the status bar when new messages arrive.

12. Touch to select the ringtone that plays when you receive new messages.

13. Touch to enable or disable vibration when new messages are received.

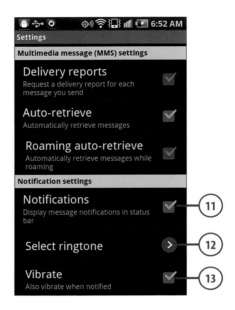

Messages Application Settings (DROID Incredible 2)

1. Press the Menu button and touch Settings.

2. Touch Notifications.

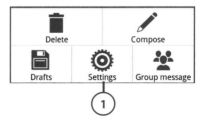

3. Touch to enable or disable whether a message is displayed in the status bar when new messages arrive.

4. Touch to enable or disable playing a sound when a new text message arrives.

5. Touch to select the ringtone that plays when you receive new messages.

6. Touch to enable or disable vibration when new messages are received.

7. Touch to enable or disable whether a message is displayed in the status bar when you send a text message.

8. Touch to enable or disable whether a message is displayed in the status bar if a text message you send fails.

9. Touch to enable or disable playing a sound when you send a text message.

10. Touch to select the ringtone that plays when you send a text message.

11. Scroll down for more settings.

12. Touch to enable or disable vibration when new messages are received.

13. Touch to select when to display an on-screen message. Your choices are Sent Success Only, Sent Failure Only, Sent Success and Failure, or No Message.

14. Press the Back button to return to the main Messages Settings screen.

15. Touch Text Messages (SMS).

16. Touch to enable or disable delivery reports for SMS or text messages that you send.

17. Touch to set the priority level of text messages you send. Your choices are Normal and High.

18. Touch to enable or disable adding your callback number to text messages you send.

19. Touch change your callback number if you enabled sending it in step 18.

20. Touch to manage messages stored on your SIM card. This is only valid for GSM DROID models or DROIDs that can roam onto GSM networks.

21. Touch to enable or disable support for special accented characters in text messages.

22. Press the Back button to return to the main Messages Settings screen.

23. Touch Multimedia Messages (MMS).

24. Touch to enable or disable delivery reports for multimedia messages you send.

25. Touch to enable or disable auto-retrieval of your messages. You probably want to leave this enabled. If you disable this then your messages will not arrive in real-time, which almost defeats the purpose of SMS and MMS.

26. Touch to enable or disable auto-retrieval of your messages when you are roaming on a carrier that is not your own. Because roaming fees can apply, you may choose to leave this disabled.

27. Touch to set the priority level of multimedia messages you send. Your choices are Normal and High.

28. Touch to set the maximum size of a multimedia message you send. This size limit also controls how much compression is done on photos and videos before they are sent to comply with the limit. Your choices are 200 kilobytes, 500 kilobytes, and 1200 kilobytes (or 1.2 megabytes).

29. Touch to view your MMS connection settings. These settings tell your DROID how to send and receive MMS messages. You cannot change them, only view them.

30. Press the Back button to return to the main Messages Settings screen.

31. Touch General.

32. Touch to enable or disable showing the sent message history when you are typing recipient names into a new message.

33. Touch to enable or disable showing the call history when you are typing recipient names into a new message.

34. Touch to enable or disable including email addresses in the search when you are typing recipient names into a new message.

35. Touch to enable or disable the deletion of messages after their limits per thread are reached. This option works in conjunction with the options shown in steps 36 and 37.

36. Touch to change the message limit, or the maximum number of text messages, per message thread. The maximum number you can type is 5000.

37. Touch to change the message limit, or the maximum number of multimedia messages, per message thread. The maximum number you can type is 999.

38. Touch to choose where to store attachments that are sent to you using text messaging. You can store them on your DROID or on the external SD card (if you have one inserted).

39. Scroll down to see more settings.

Composing Messages

When you compose a new message, you do not need to make a conscious decision whether it is an SMS (text message) or MMS (multimedia message). As soon as you attach a file to your message, your DROID automatically treats the message as an MMS. Here is how to compose and send messages.

1. Touch to compose a new message.

2. Start typing the phone number of the recipient, or if the person is in your contacts, type the name. If the name is found, touch the mobile number.

3. Touch to start typing your message.

4. Touch to send your message.

DROID 3, Pro, X2

DROID CHARGE

DROID Incredible 2

DROID 3, Pro, X2

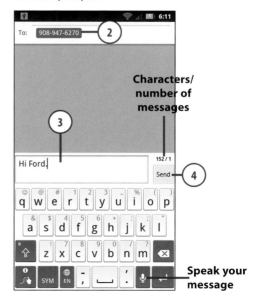

Characters/number of messages

Speak your message

DROID CHARGE

Pick from contacts, groups, or recent contacts

Speak your message

DROID Incredible 2

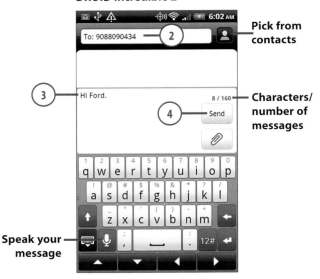

Pick from contacts

Characters/ number of messages

Speak your message

Insert Smiley Icons

To insert smiley icons, touch the Menu button and choose Insert Smiley.

Go Further

MESSAGE LIMITS AND MESSAGES

Text messages can only be 163 characters long. To get around this limit, most modern phones simply break up text messages you type into 163 character chunks. On your DROID 3, Pro, X2, and Incredible 2 you can see the number of characters you have typed, and the number of text messages your DROID will send when you hit the Send button. The phone receiving the message simply combines them all together into one message. This is important to know if your wireless plan has a text message limit. When you type and send one text message, your DROID may actually be sending two or more.

Attaching Files

If you want to send a picture, audio file, or video along with your text message, all you need to do is attach the file. Attaching a file turns your SMS message into an MMS message.

1. After you type your message but before you send your message, press the Menu button and touch Insert. On the DROID Incredible 2, just touch the attach icon.

DROID 3, Pro, X2, and CHARGE

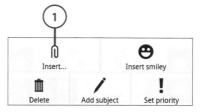

DROID Incredible 2

2. Touch to select the type of attachment.

Attachment Types

The list of attachment types varies from DROID to DROID, so the list you see here might show some attachment types that your DROID doesn't support.

How Are Attachments Shown?

Your DROID 3, Pro, and X2 do not show the attachments embedded in your message before you send it, but rather show icons on the left of the screen that represent the attachments. Your DROID CHARGE shows the attachments and has buttons to enable you to preview the attachment, replace it, or remove it. Your DROID Incredible 2 shows the attachments in the message.

3. Touch Send after you have attached all the files you want to the message.

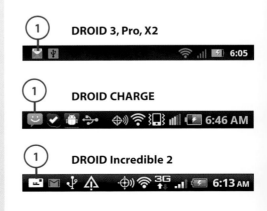

Receiving Messages

When you receive a new SMS or MMS, you can read the message, view attachments, and even save the attachments onto your DROID.

1. When a new SMS or MMS arrives you are notified with a ringtone. A notification also displays in the status bar.

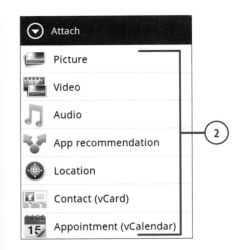

2. Pull down the status bar and touch the new message to view it.

3. Touch the text message alert to open it.

DROID 3, Pro, X2

DROID CHARGE

DROID Incredible 2

4. Touch an attachment to open it.

5. Touch and hold a message to reveal more options, which are described in the next section.

6. Touch to compose a reply to the message.

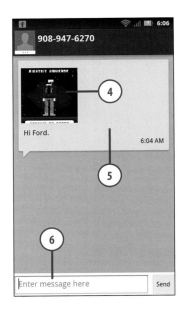

Messages Options

On your DROID 3, Pro, X2, and CHARGE, when you touch and hold a message you see options. Here are what they do.

1. Touch to prevent a message from being deleted when the thread message limit is reached. Read more about message limits in the "Messaging Application Settings" section earlier in this chapter.

2. Touch to copy the text of the message.

3. Touch to view the delivery report if the message has one.

4. Touch to forward the message.

5. Touch to view the message details.

6. Touch to delete the message.

Individual Messages versus Whole Threads

Using Delete Message in the Message Options only deletes the current message, not the entire message thread. If you want to delete the entire thread of messages, use the Delete Threads option described in the "Messaging Application Settings" section.

Usable Content

If a text message contains links to websites, phone numbers, or email addresses, touching those links results in the appropriate action. For example, touching a phone number tells your DROID to call the number, and touching a web link opens the page in the browser.

In this chapter, you learn how to set the time, use the Clock application, and use the Calendar application. Topics include the following:

→ Synchronizing to the correct time
→ Working with the Clock application
→ Setting alarms
→ Working with the Calendar

Date, Time, and Calendar

With the exception of the DROID X, your DROID has a great Clock application that you can further enhance with the use of the optional Desktop Dock. The Calendar application synchronizes to your Google or Microsoft Exchange Calendars and enables you to create meetings while on the road and to always know where your next meeting is.

Setting the Date and Time

Before we start working with the Clock and Calendar applications, we need to make sure that your DROID has the correct date and time.

1. Press the Menu button, and touch Settings.

2. Touch Date & Time.

3. Touch to enable or disable synchronizing time and date with the wireless carrier. It is best to leave this enabled as it automatically sets date, time, and time zone based on where you are travelling.

4. Touch to set the date if you choose to disable network synchronization.

5. Touch to set the time zone if you choose to disable network synchronization.

6. Touch to set the time if you choose to disable network synchronization.

7. Touch to enable or disable the use of 24-hour time format. This makes your DROID represent time without AM or PM. For example 1:00PM becomes 13:00.

8. Touch to change the way in which the date is represented. For example in the U.S. we normally write the date with the month first (12/31/2010). You can make your DROID display the date with day first (31/12/2010) or with the year first (2010/12/31).

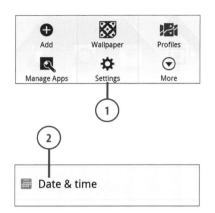

Alarm & Timer Application (DROID 3, Pro, X2)

The Clock application is designed to be used as a bedside alarm clock. However unlike a regular alarm clock, the Clock application has some special features. It can display the weather, and it enables you to watch videos or listen to music. The Clock application is very useful, especially when used in conjunction with the optional Desktop Dock.

Navigating the Alarm & Timer Application

There are two ways to launch the Clock application.

1. Touch the Alarm & Timer icon from the Launcher.

2. Touch to show the clock and alarm tab.

3. Touch to use the Timer.

4. Touch to enable or disable an alarm.

5. Touch to edit an alarm.

6. Press the Menu button to reveal options.

7. Touch to add a new alarm.

8. Touch to hide the clock and only show the alarms.

9. Touch to change the settings of the Alarm & Timer app.

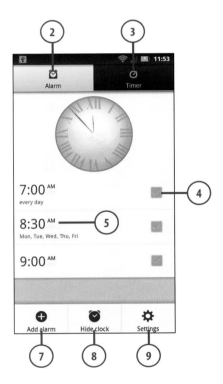

Managing Alarms

The Clock application enables you to set multiple alarms. These can be one-time alarms or recurring alarms. Even if you exit the Clock application, the alarms you set still trigger. Whether you are adding a new alarm or editing an existing one, the steps are the same.

1. Touch to edit an alarm.

2. Touch to add a new alarm.

3. Touch to enable or disable the alarm. The line turns green when an alarm is enabled.

4. Touch to change the name of the alarm. For example you can change it to "Wake Up!"

5. Touch to set the time the alarm must trigger.

6. Touch to set the sound the alarm will make when it triggers.

7. Touch to make your DROID vibrate as well when the alarm triggers.

8. Touch to choose which days of the week the alarm must trigger. This is useful if you want one alarm to wake you up at a specific time on week days but need another alarm to wake you at a different time on the weekend.

9. Touch to enable or disable the option of making your DROID continually increase the volume of the alarm until you wake up.

10. Touch to enable or disable a back-up alarm that sounds if you have not woken up from the main alarm after five minutes.

11. Touch Done.

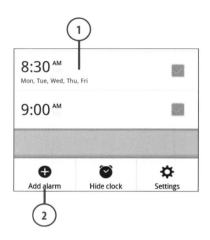

Overall Alarm Settings

Use the Settings to control how all alarms function.

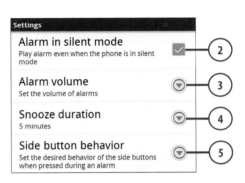

Add alarm   Hide clock   Settings

1. Press the Menu button and touch Settings.

2. Touch to enable or disable playing the alarm even if your DROID is in silent mode.

3. Touch to set the volume for all alarms.

4. Touch to set the duration of the snooze period. Your choices range between 5 and 30 minutes.

5. Touch to set how the side buttons behave if you press any of them when the alarm sounds. Your choices are Snooze and Dismiss.

Settings

Alarm in silent mode
Play alarm even when the phone is in silent mode

Alarm volume
Set the volume of alarms

Snooze duration
5 minutes

Side button behavior
Set the desired behavior of the side buttons when pressed during an alarm

Using the Timer

The Timer function can be useful for keeping track of a period of time, for example if you are cooking.

1. Touch the Timer tab.

2. Use the + and – buttons to set how many minutes the timer must run for.

3. Use the + and – buttons to set how many hours the timer must run for.

4. Touch Start.

Timer Settings

Press the Menu button and touch Settings to adjust the sound of the Timer end alarm.

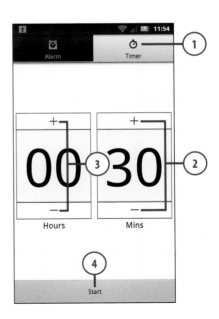

Clock Application (DROID CHARGE and Incredible 2)

The Clock application is preinstalled on your DROID CHARGE and Incredible 2 and provides functionality similar to the Alarm & Timer app on the DROID X2.

Navigating the Alarm & Timer Application

There are two ways to launch the Clock application.

1.   Touch the Clock icon from the Launcher.

Finding the Clock App on Your Incredible 2

For some reason, the Clock application isn't visible in your DROID's Launcher, even though it is installed. To find it, touch the Google search widget on the Home screen. Type clock. Touch the Clock application.

Search for clock

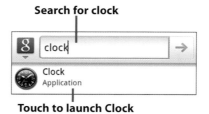

Touch to launch Clock

2.   Touch to show Alarm tab.

3.   Touch to create a new alarm.

4.   Touch to enable or disable an alarm.

5.   Touch to edit an alarm.

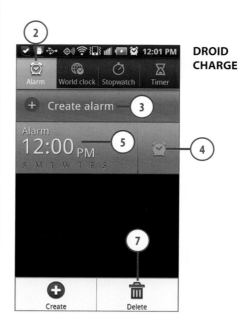

DROID CHARGE

6. Press the Menu button to reveal options.

7. Touch to delete an alarm.

8. Touch to change the overall alarm settings (DROID Incredible 2 only).

DROID Incredible 2

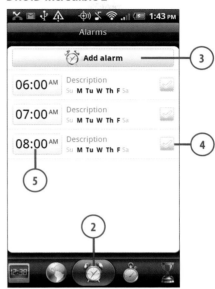

DROID Incredible 2

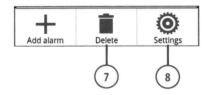

Managing Alarms (DROID CHARGE)

The Clock application enables you to set multiple alarms. These can be one-time alarms or recurring alarms. Even if you exit the Clock application, the alarms you set still trigger. Whether you are adding a new alarm or editing an existing one, the steps are the same.

1. Touch to edit an alarm.

2. Touch to add a new alarm.

3. Set the time when the alarm must trigger. Don't forget to select AM or PM.

4. Touch to choose which days of the week the alarm must trigger. This is useful if you want one alarm to wake you up at a specific time on week days but need another alarm to wake you at a different time on the weekend.

5. Touch to change the name of the alarm. For example, you can change it to "Wake Up!"

6. Touch to enable or disable the snooze function for this alarm.

7. Scroll down for more options.

8. Touch to choose how the snooze function works for this alarm. You can choose how many times the snooze repeats and how many minutes between snooze alerts.

9. Touch to enable or disable the snooze function for this alarm

10. Touch to enable or disable the feature that displays the Daily Briefing screen after you have dismissed the alarm.

Daily Briefing Screen

The Daily Briefing screen will show your weather, stocks, news, and any upcoming calendar appointments. Swipe from left to right to scroll through the screens.

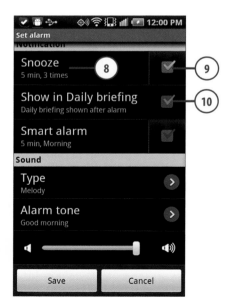

11. Touch to enable or disable the Smart Alarm.

12. Touch to choose what soothing sound and music to play, and how many minutes to start it before the alarm triggers.

What Is Smart Alarm?

The Smart Alarm is a feature on your DROID CHARGE that plays soothing sounds and music before the alarm goes off to slowly bring you out of your deep sleep.

13. Touch to select the type of alarm. You can choose Melody, vibration, or both.

14. Touch to choose the sound the alarm makes when it triggers.

15. Adjust the volume of the alarm.

16. Touch Save.

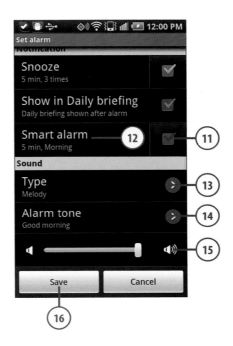

Swipe left and right to switch screens

Managing Alarms (DROID Incredible 2)

The Clock application enables you to set multiple alarms. These can be one-time alarms or recurring alarms. Even if you exit the Clock application, the alarms you set still trigger. Whether you are adding a new alarm or editing an existing one, the steps are the same.

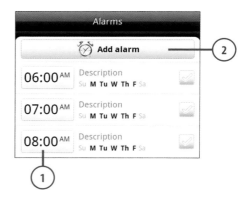

1. Touch to edit an alarm.

2. Touch to add a new alarm.

3. Set the time when the alarm must trigger. Don't forget to select AM or PM.

4. Touch to change the name of the alarm. For example you can change it to "Wake Up!"

5. Touch to choose the sound the alarm plays when it triggers.

6. Touch to choose which days of the week the alarm must trigger. This is useful if you want one alarm to wake you at a specific time on week days but need another alarm to wake you at a different time on the weekend.

7. Touch to enable or disable whether your DROID also vibrates when the alarm triggers.

8. Touch Done.

Overall Alarm Settings (DROID Incredible 2)

Use the Settings to control how all alarms function.

1. Press the Menu button and touch Settings.

2. Touch to enable or disable playing the alarm even if your DROID is in silent mode.

3. Touch to set the volume for all alarms.

4. Touch to set the duration of the snooze period. Your choices range between 5 and 30 minutes.

5. Touch to set how the side buttons behave if you press any of them when the alarm sounds. Your choices are Snooze and Dismiss.

Using the Timer

The Timer function can be useful for keeping track of a period of time, for example if you are cooking.

1. Touch the Timer tab.

2. Use keypad to set how many hours, minutes, and seconds the timer must run for.

3. Touch to choose the alarm sound that plays when the timer runs out (DROID Incredible 2 only).

4. Touch Start.

DROID CHARGE

DROID Incredible 2

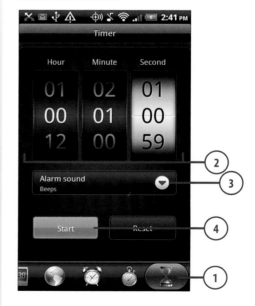

Using the Stopwatch

1. Touch the Stopwatch tab.

2. Touch Start.

DROID CHARGE

DROID Incredible 2

Using the World Clock

1. Touch the World Clock tab.

2. Touch to add a city.

3. Press the Menu button to see more options.

4. Touch to delete cities.

5. Touch to manually set Daylight Savings Time for your chosen cities (DROID CHARGE only).

6. Touch to change the sort order of the list of cities you have chosen.

7. Touch to change your DROIDs date and time settings. Please see "Setting the Date and Time" earlier in this chapter for more information (DROID Incredible 2 only).

8. Touch to choose your home city (DROID Incredible 2 only).

DROID CHARGE

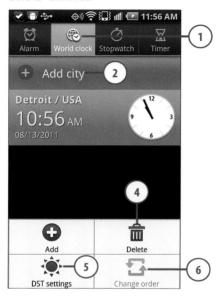

DROID Incredible 2

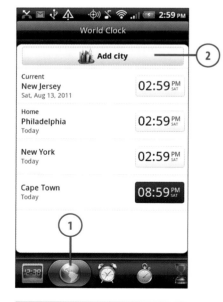

Using the Calendar Application

The Calendar application enables you to synchronize all of your Google Calendars under your primary Google account to your DROID. You can accept appointments and create and modify appointments right on your phone. Any changes are automatically synchronized wirelessly back to your Google Calendar.

The Calendar Main Screen

The main screen of the calendar shows a one-day, one-week, or one-month view of your appointments.

1. Launch the Calendar application.

2. Press the Menu button to reveal the options.

3. Touch to select the calendar to view or view all calendars.

4. Touch to see the day, week, month, or agenda (also know as list) views. On the DROID 3, Pro, and X2, this icon also enables you to jump to today's date.

5. Touch to add a new appointment.

6. Touch to see the appointments for a particular day.

7. Touch to jump to a particular date.

8. Touch to change the Calendar app settings.

DROID 3, Pro, X2

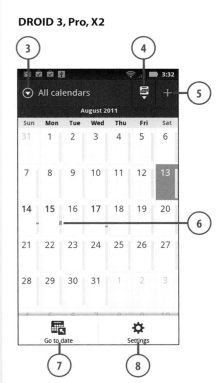

9. Touch to jump to today's date.

10. Touch to search for appointments.

11. Touch to delete appointments.

12. Touch to see more options (DROID Incredible 2 only).

Calendar Views

In the figure of Calendar's main screen we see the month view. This is where you see the entire month laid out on screen with indicators showing different appointments and where they appear in each day of the month. The week view shows the current week's seven days laid out from left to right. Appointments are visually represented during each day, but because there is more room, you can see the title for each appointment. Day view fills the screen with one day's appointments. Because there is even more room in the day view, the title and location of each appointment is visible. Agenda view displays your appointments in a list.

DROID CHARGE

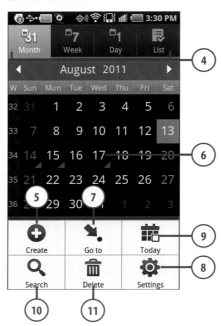

DROID Incredible 2

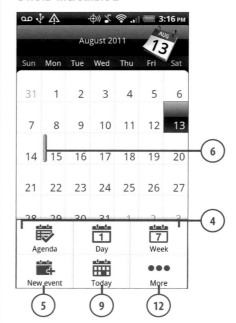

DROID Incredible 2

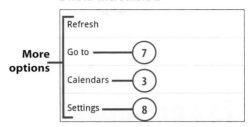

More options
- Refresh
- Go to ——— ⑦
- Calendars ——— ③
- Settings ——— ⑧

Calendar Settings (DROID 3, Pro, X2)

You can adjust how the Calendar application behaves and select which calendars you want to synchronize to your DROID.

1. From the Calendar main screen, press the Menu button and touch Settings.

2. Touch Manage Calendars to select which calendars you want to see in the calendar view and which ones you want to synchronize to your DROID.

3. Touch to enable or disable name suggestions when typing meeting attendees.

4. Touch to enable or disable Quick Conference Dialing. Quick Conference Dialing allows your DROID to automatically show your conference all information in a new meeting and enables you to select whether to include it or not.

5. Touch to manage which calendars your DROID will use Quick Conference Dialing, and enter your conference call information.

Go to date   Settings   ①

6. Touch to enable or disable hiding events that you have declined.

7. Touch to enable or disable showing meetings using your time zone only, even when you are travelling.

8. Touch to set your home time zone if you enabled home time zone in step 7.

9. Scroll down to see more settings.

10. Touch to set your week view to show only your work week, or all 7 days.

11. Touch to set how you are alerted for meeting reminders. Your choices are an alert in the status bar, an audible alert, or no reminder.

12. Touch to select the ringtone that plays when you receive a meeting reminder.

13. Touch to set whether your DROID also vibrates for meeting reminders.

14. Touch to set your default meeting reminder time.

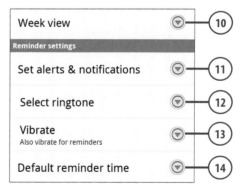

Calendar Settings (DROID CHARGE)

1. From the Calendar main screen, press the Menu button and touch Settings.

2. Touch Calendars to select which calendars you want to see in the calendar view.

3. Touch Calendar sync to change general synchronization settings such as whether to synchronize in the background or whether to automatically synchronize your calendars.

4. Touch to select which calendar view to use when you first launch the calendar app. Your choices are Month, Week, Day, and List (or agenda).

5. Touch to select how the day view is displayed. Your choices are time grid and Event list.

6. Touch to set the first day of your week. You can choose Sunday or Monday.

7. Touch to enable or disable hiding events that you have declined.

8. Touch to enable or disable showing meetings using your time zone only, even when you are travelling.

9. Scroll down to see more settings.

10. Touch to set your home time zone if you enabled home time zone in step 9.

11. Touch to set how you will be alerted for meeting reminders. Your choices are an alert in the status bar, an audible alert, or no reminder.

12. Touch to set whether your DROID also vibrates for meeting reminders.

13. Touch to select the ringtone that plays when you receive a meeting reminder.

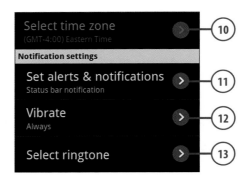

Calendar Settings (DROID Incredible 2)

1. From the Calendar main screen, press the Menu button and touch More.

2. Touch Settings

3. Touch Reminder Settings.

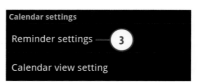

4. Touch to enable or disable apply-ing reminder settings to all calen-dars. If you enable this, the setting shown in step 5 is missing.

5. Touch to select which calendars to apply the reminder settings to. This setting is only visible if you uncheck all calendars in step 4.

6. Touch to set how you should be alerted for meeting reminders. Your choices are an alert in the status bar, an audible alert, or no reminder.

7. Touch to set the default reminder time.

8. Touch to select the ringtone that plays when you receive a meeting reminder.

9. Touch to set whether your DROID also vibrates for meeting reminders.

10. Touch to enable or disable hiding events that you have declined.

11. Press the Back button to return to the main Settings screen.

12. Touch Calendar View Settings.

13. Touch to select which calendar view to use when you first launch the calendar app. Your choices are Month, Week, Day, and Agenda.

14. Touch to select how the Day view is displayed. Your choices are Time view and Event view.

15. Touch to set the first day of your week. You can choose Sunday or Monday.

16. Touch to enable or disable includ-ing the current weather when viewing appointments.

17. Touch to select the city or state to use when generating the weather view if you enabled it in step 16.

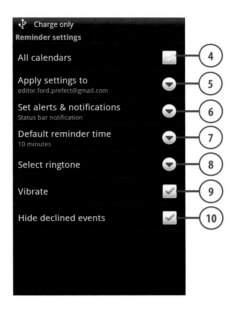

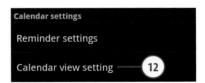

Adding a New Event/Appointment

While you're on the road you can add a new appointment or event, and even invite people to it. Events you add synchronize to your Google and corporate calendars in real-time. Depending on your model of DROID, some of the steps below are located on different parts of the screen.

1. Touch to add a new event/appointment.

2. Touch to select which calendar to add the event to.

3. Touch to enter a title for your event.

Alternative Method to Add an Event

If you touch and hold on a specific hour on a specific day, you see a pop-up window that enables you to create a new event.

4. Add event attendees. If they are already in your Contacts, their names display. If not, type the attendees' full email addresses. You can add multiple attendees separated by a comma. If your DROID is set up to synchronize with your company mailbox, you should be able to add attendees from your company's address book.

It's Not All Good

DROID Incredible 2 Strange Behavior

If you select your corporate calendar to add a new event to, the Who field is mysteriously missing. This means you cannot add any attendees to the event while you are creating it. To add attendees when creating a new event in your corporate calendar, before you save it, press the Menu button and touch Meeting Invitation. This enables you to invite attendees.

Add attendees before you save

5. Enter any optional attendees.

6. Touch to enter the date the event starts.

7. Touch to enter the time the event starts.

8. Touch to enter the date the event ends.

9. Touch to enter the time the event ends.

10. Touch to select the time zone the meeting will be held in. This is useful if you will be travelling to the meeting in a different time zone.

11. Scroll down to set more details.

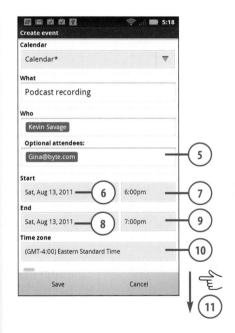

12. Touch to make this an all day event.

13. Enter where the event will be held. You can enter real addresses here.

14. Touch to include your Quick Conference information. Please see more about Quick Conference in the "Calendar Settings" section earlier in this chapter.

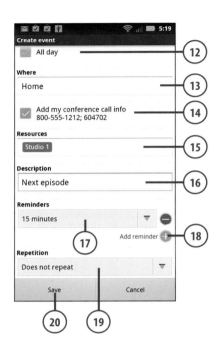

Recurring Events Are Not Flexible

Unfortunately when you choose to make an event repeat the choices you are given are not flexible. For example if you want to set up an event that repeats every Thursday, you cannot do this unless you create the event on a Thursday. Let's hope that this is addressed in a future release of Android.

15. Enter any meeting resources. This is only visible if you are adding an event to your corporate calendar.

16. Enter a description for the event.

17. Set how many minutes before the event the reminder should trigger.

18. Add an additional reminder. Additional reminders can be useful to remind yourself or others more than once of an upcoming appointment. For example, you might want to have a reminder the day before an event and a second reminder 30 minutes before the event. You can set up to five reminders in total.

19. Touch to set whether this meeting repeats, and if so, how does it repeat.

20. Touch Save. Any attendees you invited automatically receive a meeting invite via email.

It's Not All Good

Only Calendars from One Google Account

The Calendar application only synchronizes with the Google account you used to set up your DROID. Even if you add additional Google accounts in the Gmail application, the Calendar does not synchronize with them. So if you have multiple calendars under your primary Google account you should be okay, but if you use calendars from more than one Gmail account, you are not able to synchronize those additional calendars.

Edit and Delete an Event

To edit or delete a calendar event, touch and hold the event, and touch either Edit event or Delete event. When you successfully delete an event, the Calendar application sends an event decline notice to the event organizer. So you don't have to first decline the meeting before deleting it because this is all taken care of automatically.

Responding to a Gmail Event Invitation

There are actually two ways to respond to a Gmail event invitation. When you receive an email invitation in Gmail, you can either respond from the email itself or from the Calendar application.

Respond to an Event Invitation in Gmail

When someone invites you to a new event, you receive an email in your Gmail inbox with the details of that meeting.

1. Open the event invitation email in the Gmail application.

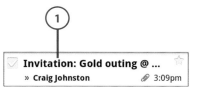

2. Touch Yes, Maybe, or No to indicate whether you will be attending.

3. Touch Use by Default for This Action. This tells your DROID that in the future when you respond to Gmail meeting invites, it must always launch the Calendar app.

4. Touch Calendar. You see the event in your Calendar view with your response already selected.

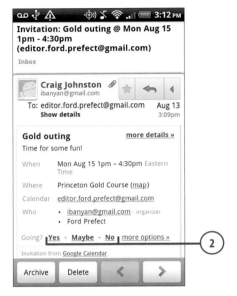

Touch to use
the Google
website instead

Respond to a Gmail Event Invitation in Calendar

When you receive a new Gmail event invitation, it is automatically inserted into your calendar even if you have not yet accepted it. It is added and set to Tentatively Accepted (Maybe).

1. Open the Calendar application and look for a new event. Touch the event to open it.

2. Select your response to the event.

3. Press the Back button to save your meeting response.

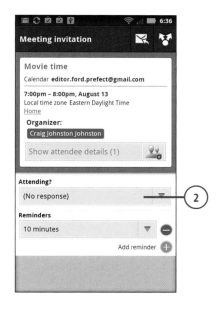

Responding to a Corporate Event Invitation

There are actually two ways to respond to a corporate event invitation. When you receive an email invitation in your corporate mailbox, you can either respond from the email itself or from the Calendar application.

Respond to an Event Invitation in Your Corporate Mailbox

When someone invites you to a new event, you receive an email in your corporate inbox with the details of that meeting.

1. Open the event invitation email.

2. Touch View Meeting.

3. Select your response to the event.

4. Press the Back button to save your meeting response.

Respond to a Gmail Event Invitation in Calendar

When you receive a new corporate event invitation, it is automatically inserted into your calendar even if you have not yet accepted it. It is added and set to Tentatively accepted (Maybe).

1. Open the Calendar application and look for a new event. Touch the event to open it.

2. Select your response to the event.

3. Press the Back button to save your meeting response.

In this chapter, you learn how to take pictures with your DROID, how to store them, and how to share them with friends. Topics include the following:

- → Using the camera
- → Sharing pictures
- → Synchronizing pictures
- → Viewing pictures

Taking, Storing, and Viewing Pictures

Your DROID has a decent 5 megapixel camera with mechanical auto-focus. This means it can take really good pictures. After you take those great pictures, you can share them with friends. You can also synchronize the pictures to your computer to possibly print out.

Using the Camera

Let's start by looking at the Camera application itself before we discuss sharing pictures with friends or synchronizing them with your computer. Remember that you must have a Micro-SD card inserted into your DROID before taking pictures.

How Much Memory Should I Buy?

A 5 megapixel picture is about 1 megabyte in size, on average. If you buy a 2 Gigabyte SD card, you could take about 2,000 pictures with the camera before the card fills up. Of course, you will be using your SD card for everything, music, video, and pictures, so you probably want to buy at least a 4 Gigabyte card or higher.

DROID Pro, X2

Let's take a tour of the DROID Pro and X2's camera.

1. Touch to launch the Camera.

2. Touch to reveal the camera menu.

3. Current location calculated by the GPS chip. If you are indoors your DROID might have a hard time determining your location.

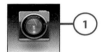

4. Touch to review pictures in the Gallery application. See more about reviewing, sharing, and editing pictures in the Gallery application for your DROID later in this chapter.

5. Touch to select different scenes for your picture. Scenes are predefined camera settings that configure it for situations such as night, portrait, sports, and more.

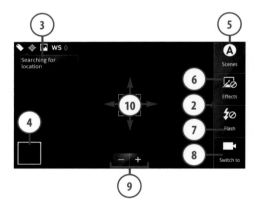

6. Touch to add effects such as negative, sepia, or black-and-white to your pictures.

7. Touch to change the flash mode from automatic to always on or always off.

8. Touch to switch between photo and video mode.

9. Use the + and − buttons to zoom in and out.

10. Drag the focus box to the area of the picture you want the camera to focus on.

11. Touch to take a picture.

Camera Modes

The DROID's camera can be set to
certain modes such as Single Shot
(that's normal mode), Panorama, and
Multi-shot. Let's take a look at each
one. To get to these camera modes,
follow these steps.

1. Press the Menu button while in
   the main camera view and then
   touch Picture Modes.

2. Touch for Single shot pictures.
   This is the normal method of taking pictures, one shot at a time.

3. Touch for Panorama mode. This
   mode helps you take panoramic
   pictures.

4. Touch for Multi-shot mode, which
   takes six pictures in quick succession.

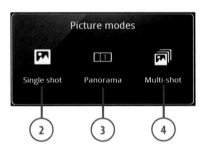

TAKE A PANORAMIC PICTURE

Your DROID's Panorama mode helps you take panoramic photos by guiding
you from one shot to the next. Each panorama consists of six pictures that
are stitched together to make one photo. Each picture in the panorama is 3
megapixels. Take the first picture in the panorama. You then see a guide
appear at the bottom left of the screen. Move the DROID slowly in the direc-
tion you want to cover the entire scene. The camera automatically takes the
next picture when it determines that you have moved the frame sufficiently.
Continue moving until the camera takes six pictures. After the sixth picture,
the camera automatically stitches the pictures together to form one large
panoramic photo.

Go Further

TAKE PICTURES IN SPURTS

Your DROID can snap six pictures instead of one when you press the camera button. The only drawback is that each picture is only 1 megapixel, but it does enable you to potentially get the right shot.

Press the Menu button and then touch Picture Modes. Touch Multi-shot. Take a picture as you normally would. Your DROID takes six pictures in quick succession.

Picture Tags

When you take pictures on your DROID, it can embed tags in the pictures. Normally it embeds the GPS coordinates and place names in the pictures, but you can add your own tags, especially if you know that GPS is not locking in.

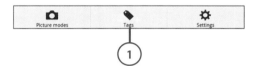

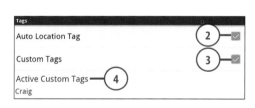

1. Press the Menu button while in the main camera view and then touch Tags.

2. Touch to disable or enable auto location tagging. Location tagging uses GPS to calculate where you are.

3. Touch to enable or disable custom tags.

4. Touch to add custom tags.

5. Enter a new tag.

6. Touch to add the new tag.

7. Touch Done.

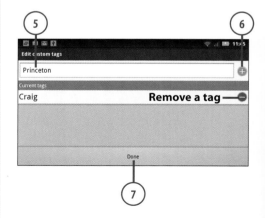

Camera Settings

Using camera settings you can change things such as the resolution of each picture, picture review time, and more.

1. Press the Menu button while in the main camera view and then touch Settings.

2. Touch to change the picture resolution. You can choose between a tiny picture of 1 megapixels to a large picture of 8 megapixels.

What Is Widescreen?

Widescreen is the default resolution setting of the camera. It takes a widescreen image that looks good on your DROID's camera screen. The issue with Widescreen images are that they are not at full camera resolution. If you want the best pictures your DROID can take, don't use Widescreen. You can always crop pictures later to suit your needs.

3. Touch to change the video resolution. Read more about video in Chapter 3, "Audio and Video."

4. Touch to set the picture exposure time. You can change it in a range from –2 to 2.

5. Touch to enable or disable the camera shutter tone. This is the camera shutter noise that plays when you take a picture to make it sound like a real camera.

6. Scroll down to see more options.

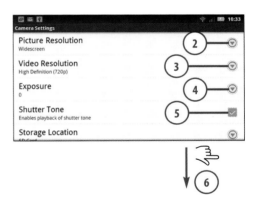

7. Touch to choose where to save pictures take with the camera app. You can use your internal phone storage or an SD card (if you have one inserted).

8. Touch to change the focus options. Your choices are Continuous Focus, which makes the camera continually refocus, and Auto Focus, which only focuses when you touch and hold the shutter icon. Continuous Focus does use more battery than Auto Focus.

DROID 3

1. Touch to launch the Camera.

2. Press the Menu button to reveal the bottom slide-out menu.

3. Current location calculated by the GPS chip. If you are indoors your DROID might have a hard time determining your location.

4. Touch to review pictures in the Gallery application. See more about reviewing, sharing, and editing pictures in the Gallery application for your DROID later in this chapter.

5. Touch to select different scenes for your picture. Scenes are predefined camera settings that configure it for situations such as night, portrait, sports, and more.

6. Touch to add effects such as negative, sepia, or black-and-white to your pictures.

7. Touch to change the flash mode from automatic to always on or always off.

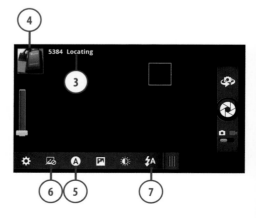

8. Touch to switch between photo and video mode.

9. Use the slider to zoom in and out.

10. Drag the focus box to the area of the picture you want the camera to focus on.

11. Touch to take a picture.

12. Touch to switch between the front and back cameras.

13. Touch to choose between panoramic or single-shot camera modes.

14. Touch to change the brightness.

15. Touch to change the camera settings.

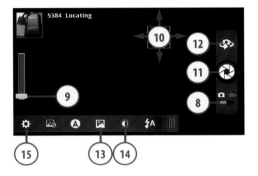

Camera Settings

1. Press the Menu button while in the main camera view and then touch Settings.

2. Touch to change the picture resolution. You can choose between a tiny picture of 1 megapixels, to a large picture of 8 megapixels.

3. Touch to change the video resolution. Read more about video in Chapter 3.

4. Touch to choose where to save pictures take with the camera app. You can use your internal phone storage or an SD card (if you have one inserted).

5. Touch to enable or disable whether to tag your pictures with your current location.

6. Scroll down to see more options.

7. Touch to enable or disable the shutter sound that plays when you take a picture.

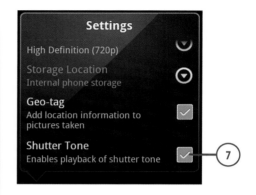

DROID CHARGE

1. Touch to launch the Camera.

2. Touch to reveal the camera menu.

3. Indicates how many pictures you can take at the current resolution based on how much memory is available.

4. Touch to review pictures in the Gallery application. See more about reviewing, sharing, and editing pictures in the Gallery application for the DROID later in this chapter.

5. Touch to take a picture.

6. Touch to switch between photo and video mode.

7. Touch to switch between the front and back cameras.

8. Touch to select different shooting modes. These include modes that allow you take panoramic pictures and action shots.

9. Touch to change the flash mode from automatic to always on or always off.

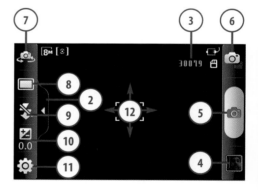

10. Touch to change the camera exposure settings.

11. Touch to change the camera settings.

12. Drag the focus box to the area of the picture you want the camera to focus on.

Camera Settings

1. Pull out the left menu while in the main camera view and then touch Settings.

2. Touch for picture settings.

3. Touch to enhance pictures when outdoors.

4. Touch to set the Focus Mode. You can choose between Auto Focus, Macro (taking pictures of objects close up), and Face Detection, which focuses on people's faces in a picture.

5. Touch to select the scene mode. This enables you to choose different scenes such as landscape, portrait, sports, and party.

6. Touch to go to the next screen of settings.

7. Touch to choose whether to use a timer when taking pictures. You can choose to make the camera wait up to 10 seconds before snapping the picture, which is good if you want to prop your DROID up and get into the shot.

8. Touch to choose the resolution of the pictures being taken. Higher resolution means crisper, better pictures with large file sizes.

9. Touch to set the white balance or leave it on automatic. You can manually choose levels for daylight, cloudy days, and so on.

10. Touch to go to the next screen of settings.

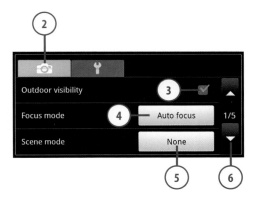

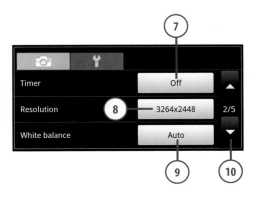

11. Touch to apply effects to the pictures you are taking. You can choose effects such as sepia, negative, and black and white.

12. Touch to choose the ISO rating for the pictures. You can select between 100 to 800 or leave it set to automatic.

13. Touch to set the light metering for the pictures. You can set it to Centre-weighted, Spot, or Matrix.

14. Touch to go to the next screen of settings.

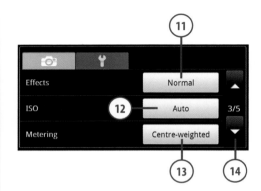

15. Touch to enable or disable the anti-shake feature that tries to compensate for camera shake.

16. Touch to enable or disable automatic contrast adjustment.

17. Touch to enable or disable blink detection, which watches for people blinking when you take a picture and alerting you so you can take another one.

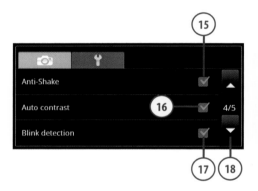

18. Touch to go to the next screen of settings.

19. Touch to adjust the image quality of the pictures after you take them. This setting controls how much compression is done on the pictures before they are stored.

20. Touch to adjust the Contrast, Color Saturation, and Sharpness of the pictures.

21. Touch to go to the camera settings.

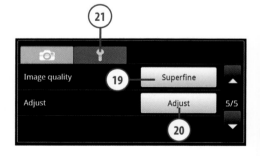

22. Touch to enable or disable showing a grid pattern on the camera screen.

23. Touch to enable or disable a picture review. This shows the picture after you have taken it before showing the view finder image.

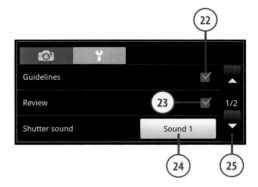

24. Touch to select the sound that the camera plays as the shutter sound or select off.

25. Touch to go to the next screen of settings.

26. Touch to enable or disable adding your GPS location to each picture you take.

27. Touch to reset all the camera settings back to factory default.

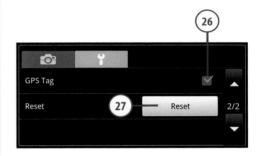

DROID Incredible 2

1. From the Home screen, touch the Camera application icon to launch the camera.

2. Touch to switch between the front and back cameras.

3. Slide to change the zoom level.

4. Touch to switch to the video recording mode. See more about video recording in Chapter 3.

5. Touch to change the flash mode from automatic to always on or always off.

6. Touch to take a picture.

7. Touch to choose camera effects to apply to your pictures as you take them. You can choose from effects such as distortion, polarize, and many more.

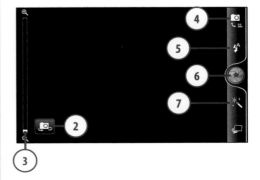

8. Touch to review, share, and delete pictures in the Photos application. See more about the Photos application later in this chapter.

9. Touch to focus on an area of the picture.

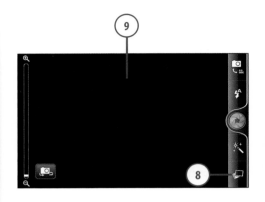

Focus on Parts of a Picture

With the DROID Incredible camera, you can actually focus on a certain part of a picture. By touching the part of the picture you want in focus, you will see that the rest of the picture goes out of focus. Using this trick enables you to take some amazing pictures.

Camera Settings

When you touch the camera settings arrow, you can change all the camera settings and even make use of effects.

1. Press the Menu button to see the camera settings.

2. Touch to use a timer when taking pictures. You can choose Off, 2 seconds, or 10 seconds.

3. Touch to change the Exposure, Contrast, Color Saturation, and Sharpness of the pictures.

4. Touch to change the white balance setting for the picture. You can leave it on automatic or choose from settings such as daylight, cloudy, and so on.

5. Touch to choose the resolution for the picture.

6. Touch to adjust ISO for the picture. You can leave it at automatic or choose between 100 and 800 ISO.

7. Scroll down for more settings.

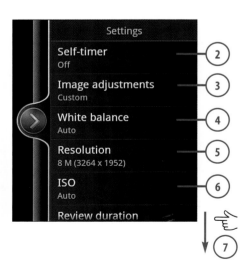

8. Touch to choose the picture review duration. This controls whether the camera shows the picture you have just taken for review.

9. Touch to switch between 4:3 ratio (standard) and 5:3 ratio (widescreen) pictures.

10. Touch to tag your picture with your current location.

11. Touch to enable or disable automatically enhancing your pictures.

12. Touch to enable or disable auto focus.

13. Scroll down for more settings.

14. Touch to enable or disable face detection. This feature recognizes and focuses on faces.

15. Touch to enable or disable playing a shutter sound.

16. Touch to draw a grid on the camera screen.

17. Touch to reset all camera settings to the factory default.

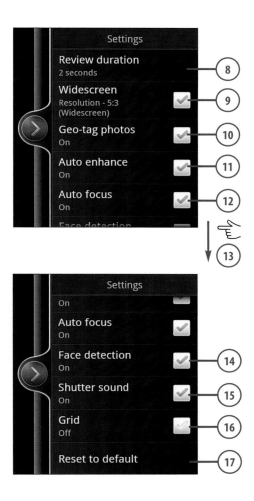

Viewing and Managing Your Photos

No matter if you have snapped pictures using your DROID, or have synchronized photos from a computer, you can use the Gallery application to manage, edit, and share your photos.

DROID CHARGE

The Gallery application on the DROID CHARGE is the standard Android gallery app.

1. Touch to launch the Gallery application.

2. Touch a thumbnail photo to open an album. Touch and hold on an album to reveal the album menu.

3. Touch the camera icon to launch the Camera application.

4. Swipe left and right to see all photo albums.

5. Touch the photos labeled as Camera to see pictures taken on your DROID.

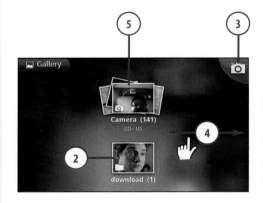

The Tilt Trick

While on the main Gallery screen, to see photos lower in the album stack, tilt your DROID forward and backward or side to side. As you tilt, you notice that the view tilts with you, allowing you to see more of the photos that are lower in the stack.

Album Menu

When you touch and hold on an album, the album menu appears. The menu enables you to share the album, delete it, or see its details.

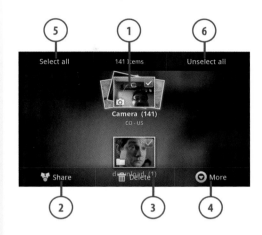

1. Touch and hold on an album to see the album.

2. Touch to share the photo album. You can share the album on Picasa, or send it via Bluetooth, Gmail, Email, or MMS.

3. Touch to delete the photo album.

4. Touch to see the photo's filename.

5. Touch to select all photo albums.

6. Touch to deselect all photo albums.

No Facebook Album Sharing
Strangely, you cannot share a photo album on Facebook. You can share photos with Picasa, though. Let's hope that a future version of Android will allow sharing with Facebook.

Managing and Sharing Photos in an Album

After you open a photo album, you can manage the photos in it, edit them, and share them.

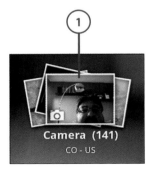

1. Touch an album to see all pictures in it.

2. Slide to switch between stacked and grid views.

3. Slide left and right to rapidly move through the pictures.

4. Touch a picture to open it.

Name of the album

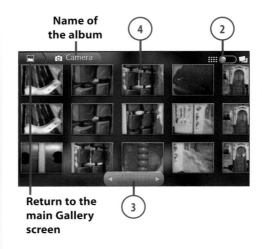

Return to the main Gallery screen

Review, Edit, and Share Pictures

After you open a picture you can review it, edit it, and share it.

1. Touch the picture to reveal the menus.

2. Touch to view a slideshow of all pictures in the album.

3. Touch to zoom in and out.

4. Swipe left and right to scroll between the pictures in the album.

5. Touch to see more options such as sharing pictures, cropping them, and using them as wallpaper.

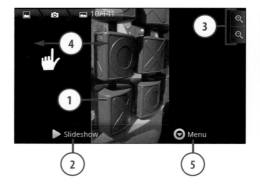

Rotating, Cropping, and Mapping Pictures

If you press the Menu button and then touch More while you're looking at a photo in the picture review screen (step 5 in the "Review, Edit, and Share Pictures" section), you can crop the picture, rotate it, and see it on a map.

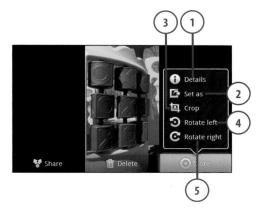

1. Touch to see the details about the picture, such as the size of it on the memory card, the filename, the date it was taken, and the album it is in. The details might include META data such as the GPS coordinates if they were originally inserted.

2. Touch to set the picture as your wallpaper or contact icon.

3. Touch to crop the picture.

4. Touch to rotate the picture left.

5. Touch to rotate the picture right.

It's Not All Good

Cannot Create an Album

While you can edit, delete, and share a photo album (as described in the "Album Menu" section), you cannot create a new album. The only way to create a new photo album is on your desktop computer and synchronize it to your DROID.

DROID Incredible 2

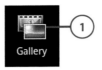

Gallery

1. Touch to launch the Gallery app.

2. Touch to see photo albums on your DROID.

3. Touch to see your Facebook photo albums and those of your friends on Facebook.

4. Touch to see photo albums on Flickr.

5. Touch to view photos on devices that support DLNA (Digital Living Network Alliance). These could be satellite receivers, network connected hard disk, and more.

6. Touch an album to open it.

Review, Edit, and Share Pictures in Album View

When you open the album view, you can review, edit, and share pictures.

1. Touch an album in the main Photos screen to see the pictures in the album.

2. Touch to return to the main Gallery screen.

3. Touch to share the picture with people via Bluetooth, Facebook, Flickr, Gmail, MMS, and more.

4. Touch to select one or more pictures to delete.

5. Touch to launch the Camera app.

6. Swipe up and down to see all photos.

7. Touch a photo to see it full screen.

8. Touch and hold a picture to see more options.

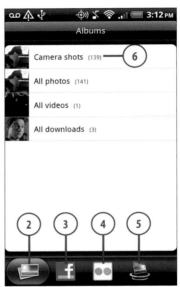

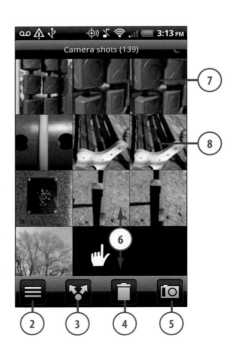

9. Touch to view the picture full screen.

10. Touch to edit the picture by cropping it, rotating it left or right, or applying effects to it.

11. Touch to share the picture with friends via Facebook, MMS, email, and more.

12. Touch to use the picture as a contact icon, set it as one of your favorites, or as your DROID's wallpaper.

13. Touch to delete the picture.

14. Touch to start a slideshow of all the pictures in the album.

15. Touch to see details about the picture.

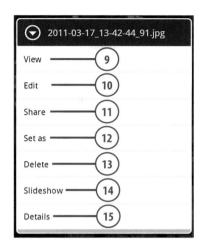

DROID 3 and DROID X

1. Touch to launch the Gallery application.

2. Scroll left and right to see all pictures from your friends on Facebook.

3. Touch to see pictures in your camera roll.

4. Touch to see pictures in your photo library.

5. Touch to see pictures on photo sharing sites such as Picasa and social networking sites such as Facebook.

6. Touch to see pictures from your friends on social networking sites.

7. Touch to see pictures on devices that support DLNA (Digital Living Network Alliance). These can be satellite receivers, networking connected hard disks, and more.

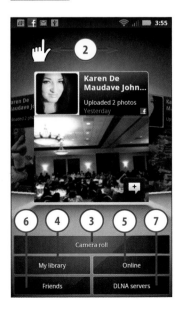

Settings

The Gallery application has a few settings such as the slideshow interval, slideshow transition, and more. Let's take a look.

1. Press the Menu button and touch Settings.

2. Touch to change the slideshow interval to 2, 3, or 4 seconds.

3. Touch to enable or disable repeating slideshows.

4. Touch to enable or disable shuffling slides when displaying a slide show.

5. Touch to set the transition between slides when watching a slide show. Your choices include Fade, Shrink and Grow, and more.

6. Touch to set an account to use when choosing the Quick Upload feature. Quick Upload enables you to quickly upload pictures to a preset photo sharing site.

7. Touch to set an account to use when choosing the video Quick Upload feature. Quick Upload allows you to quickly upload video to a preset video sharing site.

8. Touch to set the DLNA Media Share settings. Media Share enables you to share your pictures and video with devices that support DLNA.

9. Touch to enable or disable autocompleting recognized names in your Contacts when adding tags to pictures.

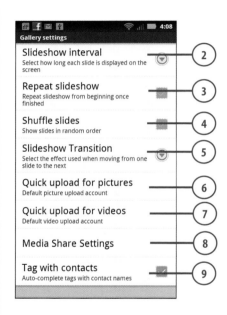

Reviewing, Sharing, and Editing Pictures

When you touch the camera roll or photo library album, you see all pictures in that album. From there you can share them, edit them, or simply view them. Let's take a look.

1. Press the Menu button to see album options.

2. Touch to launch the Camera app.

3. Touch to start a slideshow of all pictures in the album.

4. Touch to return to the Gallery main screen.

5. Touch to change the way the pictures are grouped on the screen. You can group them by day, week, and month.

6. Touch to select multiple items. This enables you to select multiple items and then press the Menu button to share them, tag them, delete them, or print them.

7. Touch to go to the Gallery settings. See the "Camera Settings" section earlier in this chapter for more information.

8. Touch to open a picture full screen.

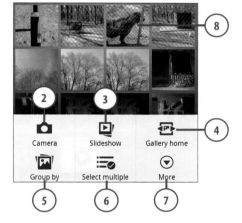

Reviewing and Sharing Pictures

When you touch a picture in an album you can view it, edit it, share it, and even print it.

1. Touch the picture after it opens full screen to see a short menu.

2. Shows the location where the picture was taken (if known).

3. Shows the picture information including the filename and date it was taken.

4. Touch to share the picture with friends via Bluetooth, Gmail, Picasa, MMS, and many other options.

5. Touch to use the Quick Upload feature to quickly upload the picture to your preset Quick Upload album. See the Settings section earlier in the chapter to learn how to set the Quick Upload album.

6. Touch to comment on the picture. This only works for pictures from social networking sites.

7. Touch to launch the Camera app.

8. Press the Menu button to see more options.

9. Touch to return to the Gallery main screen.

10. Touch or edit the picture. This includes rotating it, cropping it, changing the picture tags, and advanced editing.

11. Touch to delete the picture.

12. Touch to add the picture to an album.

13. Touch to set the picture as a contact icon, your Facebook profile picture, or your DROID's wallpaper.

14. Touch to see more options.

15. Touch to show on a map where the picture was taken. This only works if the picture has location information in it.

16. Touch to start a slideshow of all the pictures in the album.

17. Touch to copy the picture to a media server that supports DLNA.

18. Touch to display the picture on a device that supports DLNA and is connected to a television.

19. Touch to print the picture at a physical store (such as CVS) or to a Bluetooth printer.

20. Touch to see the picture information.

21. Touch to change the Gallery app settings.

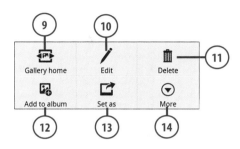

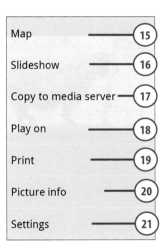

DROID Pro

1. Touch to launch the Gallery application.

2. Touch to launch the Camera app.

3. Touch an album to see the picture within it.

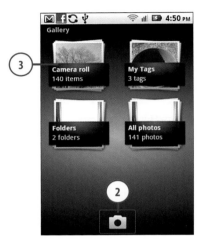

Settings

The Gallery application has a few settings such as the slideshow interval, slideshow transition, and more.

1. Press the Menu button and touch Settings.

2. Touch to change the slideshow interval to 2, 3, or 4 seconds.

3. Touch to enable or disable repeating slideshows.

4. Touch to enable or disable shuffling slides when displaying a slide show.

5. Touch to set the transition between slides when watching a slide show. Your choices include Fade, Shrink and Grow, and more.

6. Touch to set an account to use when choosing the Quick Upload feature. Quick Upload enables you to quickly upload pictures to a preset photo sharing site.

7. Touch to enable or disable auto-completing recognized names in your Contacts when adding tags to pictures.

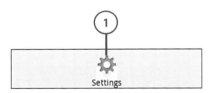

Reviewing, Sharing, and Editing Pictures

When you touch the camera roll or photo library album, you see all pictures in that album. From there you can share them, edit them, or simply view them. Let's take a look.

1. Press the Menu button to see album options.

2. Touch to launch the Camera app.

3. Touch to start a slideshow of all pictures in the album.

4. Touch to see pictures taken on a specific date.

5. Touch to select multiple items. This enables you to select multiple items and then press the Menu button to share them, tag them, delete them, or print them.

6. Touch to go to the Gallery settings. See the "Camera Settings" section earlier in this chapter for more information.

7. Touch to open a picture full screen.

8. Touch and hold a picture to see picture options.

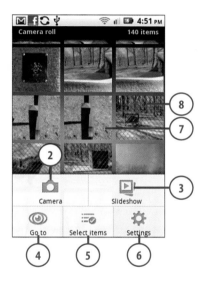

9. Touch to upload the picture to your preset Quick Upload album.

10. Touch to share the picture with friends using email, Bluetooth, or on sites such as Facebook.

11. Touch to edit the picture. Editing includes adding or changing the tags, rotating the picture, or more advanced editing.

12. Touch to delete the picture.

13. Touch to set the picture as your Facebook profile picture, a contact picture, or your DROID's wallpaper.

14. Touch to print the picture to a Bluetooth printer or to a physical store such as CVS.

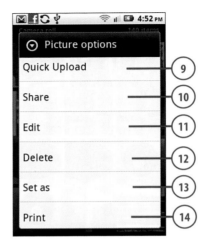

Reviewing and Sharing Pictures

When you touch a picture in an album you can view it, edit it, share it, and even print it.

1. Touch the picture when it opens full screen to see a short menu.

2. Touch to launch the Camera app.

3. Shows the date the picture was taken.

4. Touch to share the picture with friends via Bluetooth, Gmail, Picasa, MMS, and many other options.

5. Touch to use the Quick Upload feature to quickly upload the picture to your preset Quick Upload album. See the "Camera Settings" section earlier in the chapter to learn how to set the Quick Upload album.

6. Touch to see more options.

7. Touch to delete the picture.

8. Touch to set the picture as your Facebook profile picture, a contact picture, or use it as your DROID's wallpaper.

9. Touch to edit the picture, which includes changing the tags, rotating it, or more advanced editing.

10. Touch to print the picture to a Bluetooth printer or to a physical store such as CVS.

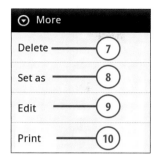

Synchronizing Photos with Your Computer

When you connect your DROID to a computer, you can move pictures back and forth manually, or by using software such as The Missing Sync or doubleTwist. This section covers how to synchronize manually and how to do it using doubleTwist. If you have not yet installed doubleTwist, follow the installation steps in the Prologue.

Working with Pictures Manually

The most generic way of working with photos and photo albums is to mount your DROID as a drive on your computer.

1. Plug your DROID into your computer using the supplied USB cable.

2. Pull down the status bar to reveal the USB Connection notification.

3. Touch the USB Connection notification.

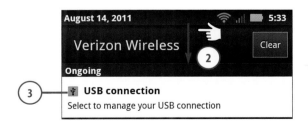

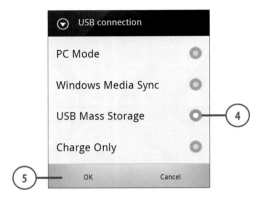

4. Touch USB Mass Storage.

5. Touch OK, and your DROID's external memory (the Micro-SD card) is mounted as a drive on your computer. If you have a Mac, the drive is called DROID, NO NAME or Untitled 1. If you have Windows, the drive is called Removable Disk (X:).

Moving Pictures (Mac OSX)

After your DROID is connected to your Mac and mounted, you can browse the DROID just like any other drive on your computer. The pictures are in a folder called DCIM.

1. Browse to your DROID to locate the pictures.

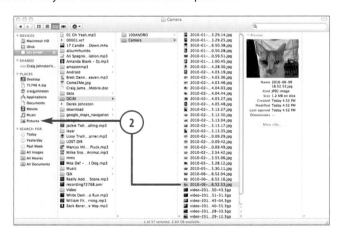

2. Drag one or more pictures from your DROID to a folder on your Mac.

Where Are the Pictures?

Pictures taken with the DROID's camera are in DCIM\Camera. All other pictures are in a folder called Images.

3. When you are done, remember to right-click on the drive and choose Eject. This is crucial to preserving the contents of your DROID.

Moving Pictures (Windows)

After your DROID is connected to your Windows computer and mounted, you can browse the DROID just like any other drive on your computer. The pictures are in a folder called DCIM.

1. When the AutoPlay window appears, click Open Folder to View Files.

Where Are the Pictures?

Pictures taken with the DROID's camera are in DCIM\Camera. All other pictures are in a folder called Images.

2. Drag one or more pictures from your DROID to a folder on your Windows computer.

3. When you are done, remember to right-click on the drive and choose Eject. This is crucial to preserving the contents of your DROID.

Creating Photo Albums

There is currently no way to create photo albums on your DROID or in doubleTwist. To manually create photo albums, browse your DROID after it is connected to a computer. Create a new folder called Images, if it doesn't already exist. Under the folder Images, create a new folder for each new photo album you want to create. The name of the folder becomes the name of the photo album. Copy photos to your new folders. After you eject the DROID, your new photo albums are visible in the Gallery application.

Automated Picture Importing (Mac OSX)

When you mount your DROID as a drive on a Mac, iPhoto normally launches. This is because iPhoto sees the DROID as a digital camera.

1. Click on the camera named NO NAME. This is your DROID.

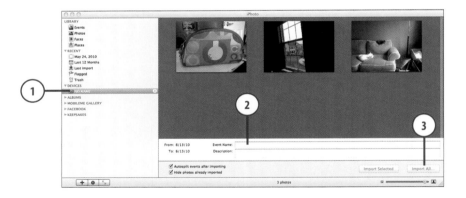

2. Enter details of the pictures.

3. Click Import All.

Automated Picture Importing (Windows)

When you mount your DROID as a drive on a Windows computer, the AutoPlay window always appears.

1. Click Import Pictures.

2. Enter one or more tags for the pictures. A tag is a key word.

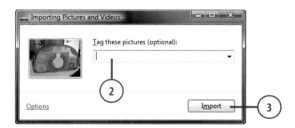

3. Click Import.

Working with doubleTwist

doubleTwist is a free download that enables you to synchronize media to and from your DROID. If you have not yet installed doubleTwist, follow the instructions in the Prologue. The steps described here are the same for Windows and Mac OSX.

1. Connect and mount your DROID so that it displays under Devices in doubleTwist.

2. Click Photos.

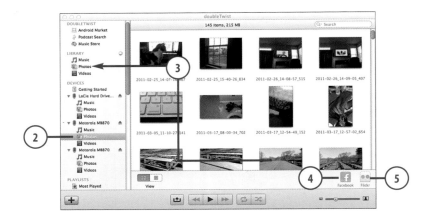

3. Drag one or more pictures from your DROID to the Photos folder in the doubleTwist Library. Your pictures are physically stored on your computer in a folder called doubleTwist, which is within the Pictures folder.

4. Click to send one or more pictures to Facebook.

5. Click to send one or more pictures to Flickr.

6. To copy pictures from your doubleTwist library to your DROID, click on Photos and drag one or more pictures to Photos on your DROID.

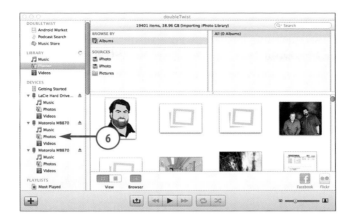

Configure doubleTwist Photo Options (Mac OSX)

Your doubleTwist Photo Library can be configured to show one or more folders on your computer where you normally store pictures. On a Mac, your iPhoto library is automatically a part of the doubleTwist library.

1. Click doubleTwist, Preferences.

2. Click Library.

3. Click to add new folders.

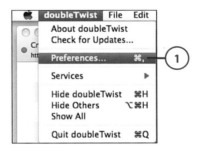

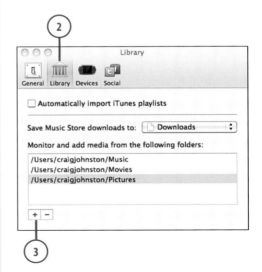

Configure doubleTwist Photo Options (Windows)

Your doubleTwist Photo Library can be configured to show one or more folders on your computer where you normally store pictures.

1. Click Edit, Preferences.

2. Select the Library tab.

3. Click a folder button to add new folders.

4. Click to save.

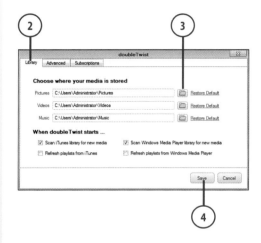

In this chapter, you learn how to purchase and use Android applications on your DROID. Topics include the following:

→ Finding applications with Android Market
→ Purchasing applications
→ Keeping applications up to date

10

Working with Android Applications

Your DROID comes with enough applications to make it a worthy smartphone. However, wouldn't it be great to play games, update your Facebook and Twitter statuses, or even keep a grocery list? Well, finding these types of applications is what the Android Market is for. Read on to learn about finding, purchasing, and maintaining applications.

Configuring Google Checkout

Before you start buying applications in the Android Market, you must first sign up for a Google Checkout account. If you plan to only download free applications, you do not need a Google Checkout account.

1. From a desktop computer or your DROID, open the web browser and go to http://checkout.google.com.

2. Sign in using the Google account that you will be using to synchronize email to your DROID. See Chapter 1, "Contacts," or Chapter 5, "Emailing," for information about adding a Google account to your DROID.

3. Choose your location. If your country is not listed, you have to use free applications until it's added to the list.

4. Enter your credit card number. This can also be a debit card that includes a Visa or MasterCard logo, also known as a check card, so that the funds actually are withdrawn from your checking account.

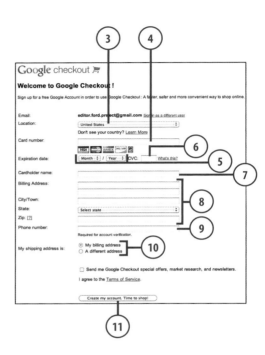

5. Select the month and year of the card's expiration date.

6. Enter the card's CVC number, which is also known as the security code. This is a three- or four-digit number that's printed on the back of your card.

7. Enter your name.

8. Enter your billing address.

9. Enter your phone number.

10. Although you don't need a mailing address for the Android Market, you might want to choose an alternative delivery address for items your purchase from online stores that use Google Checkout.

11. Click Create My Account when you're done with the form.

Navigating Android Market

Android is the operating system that runs your DROID and, therefore, any applications that are made for your DROID need to run on Android. The Android Market is a place where you can search for and buy Android applications.

1. Touch the Market application icon on the Home screen.

2. Press the Menu button to show all options.

3. Touch Apps to browse all non-game applications.

4. Touch Games to browse only games.

5. Touch Verizon to browse apps provided by Verizon. If you are using a non-U.S. version of the phone, this option might not be available.

6. Swipe left and right to scroll through the featured apps.

7. Touch an app to see more information about it.

8. Touch to search for applications.

9. Touch My Apps to see applications that you have previously downloaded. This menu item is sometimes called Downloads instead of My Apps.

10. Touch Settings to change the settings for the Market app.

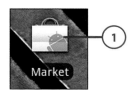

Browsing Apps by Category

If you know the type of application you want to buy or download, but want to see what is available, you can browse by category. The following steps work the same for games or other applications.

1. Touch either Apps or Games (in this example, am browsing Apps).

2. Touch a category. (In this example, I use Social.)

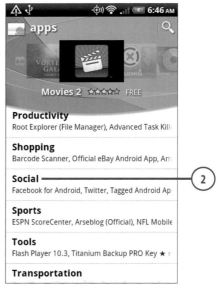

3. Touch to see the top paid applications. This is a list of most popular applications that are not free.

4. Touch to see the top free applications. This is a list of most popular applications that are free.

5. Touch Just In to see what applications have just been added to the Android Market.

6. Touch to search for an application in this category.

7. Touch an application to buy it or download it, if it is free.

Downloading Free Applications

You don't have to spend money to get quality applications. Some of the best applications are actually free.

1. Browse through the applications or games and touch the application you want to download.

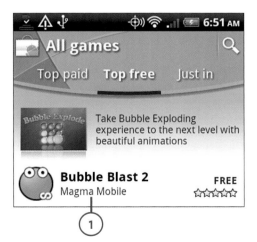

2. Scroll down to read the application features, screen shots of the app, reviews by other people who installed it, and information on the person or company who wrote the application.

3. Touch More to read more of the app description.

4. Touch FREE to download the app.

5. Touch OK to accept the app permissions and proceed with the download.

Beware of Permissions

Each time you download a free app or purchase an app from the Android Market, you are prompted to accept the app permissions. App permissions enable the app to use features and functions on your DROID, such as access to the wireless network or access to your phone log. Pay close attention to the kinds of permissions each app is requesting and make sure they are appropriate for the type of functionality that the app provides. For example, an app that tests network speed will likely ask for permission to access your wireless network, but if it also asks to access your list of contacts, it might mean that the app is malware and just wants to steal your contacts.

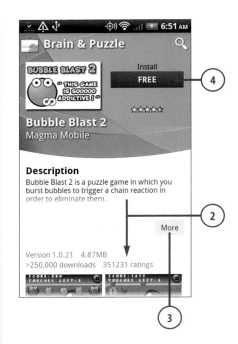

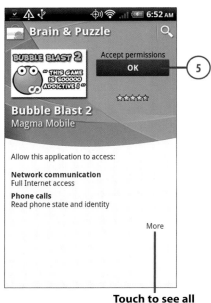

Touch to see all permissions

Buying Applications

If an application is not free, the price is displayed next to the application icon. If you want to buy the application, remember that you need to already have a Google Checkout account. See the "Configuring Google Checkout" section earlier in the chapter for more information.

1. Touch the application you want to buy.

What If the Currency Is Different?

When you browse applications in the Android Market, you might see applications that have prices in foreign currencies, such as in euros. When you purchase an application, the currency is simply converted into your local currency using the exchange rate at the time of purchase.

2. Scroll down to read the application features, app screen shots, reviews by other people who installed it, and information on the person or company who wrote the application.

3. Touch the app price to buy the application.

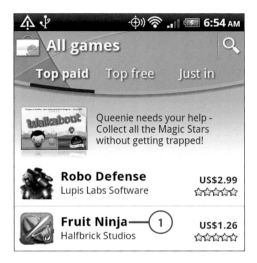

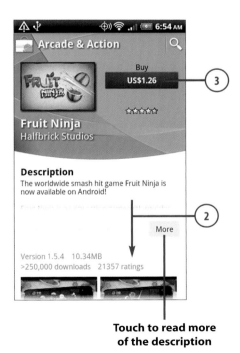

Touch to read more of the description

4. Touch OK to accept the app permissions.

5. Touch the Pay for This With dropdown to select a method of payment.

6. Touch Buy Now to purchase the app. You will receive an email from the Android Market after you purchase an app. The email serves as your invoice.

Touch to see all permissions

Deleting Applications

If you no longer have a need for a specific application, you can delete it from your DROID. There are actually two ways of doing this.

Deleting an Application Using Settings

One way you can delete applications is to use the Manage Applications menu under the Settings menu. Here is how.

1. From the Home screen, press the Menu button and touch Settings.

2. Touch Applications.

3. Touch Manage Applications.

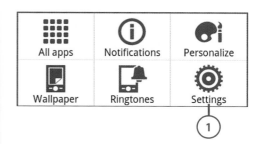

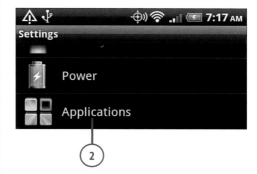

4. Touch Downloaded Applications.

5. Touch the application you want to delete.

6. Touch Uninstall.

Touch to move the app between the phone memory and the SD card

7. Touch OK to confirm that you want to delete the application.

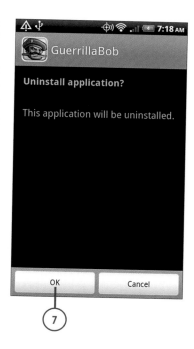

Deleting an Application Using Market

The other way to delete applications is by using the Market application. With this method, you can only delete applications that you have downloaded using the Market application.

1. From the Home screen, launch the Market application.

2. From the Market application's main screen, touch the Menu button and touch My Apps (sometimes called Downloads).

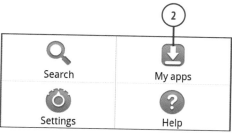

3. Touch the application that you want to delete.

4. Touch Uninstall.

5. Touch OK when the warning message is displayed.

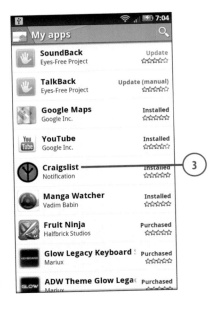

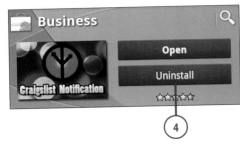

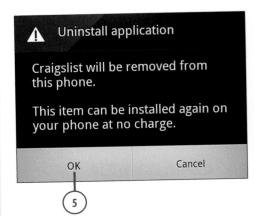

6. Select a reason why you are removing the application, or choose I'd Rather Not Say.

7. Touch OK to uninstall the application.

Accidentally Uninstall an Application?

What if you accidentally uninstall an application, or you uninstalled an application in the past but now decide you'd like to use it again? To get the application back, go to the My Apps (Downloads) view in Android Market. Scroll to that application and touch it. Touch Install to re-install it.

Deleting from the Launcher—DROID X2 Only

Your DROID X2 offers a unique way to delete an application which is actually quicker than the other methods. It enables you to delete an app right from the Launcher.

1. From the Home screen, touch the Launcher icon.

2. Touch and hold the application you want to delete.

3. Touch Uninstall.

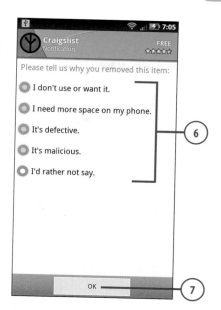

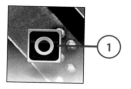

4. Touch OK.

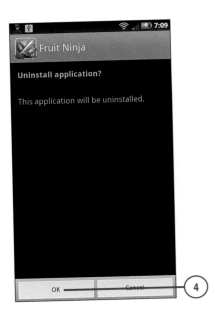

Keeping Applications Up to Date

Developers who write Android applications often update their applications to fix bugs or to add new features. With a few quick touches it is easy for you to update the applications that you have installed.

1. If an application you have installed has an update, you see the update notification in the notification bar.

2. Pull down the notification bar.

3. Touch the update notification.

4. Touch one of the applications that has an update available.

5. Touch Update.

In this chapter, you learn how to customize your DROID to suit your needs and lifestyle. Topics include the following:

→ Wallpapers and live wallpapers
→ Using Scenes
→ Replacing the keyboard
→ Sound and display settings
→ Setting region and language

Customizing Your DROID

Your DROID arrives preconfigured to appeal to most buyers; however, you might want to change the way some of the features work, or even personalize it to fit your mood or lifestyle. Luckily your DROID is customizable.

Change Your Wallpaper

Your DROID comes preloaded with a cool wallpaper. You can install other wallpapers, use live wallpapers that animate, and even use pictures in the Gallery application as your wallpaper. We start by selecting a static wallpaper.

1. Touch or press the Menu button while on the Home screen.

2. Touch Wallpaper.

3. Touch Wallpapers.

4. Touch a wallpaper to select and view it.

5. Touch Set Wallpaper after you have found a wallpaper you like. After you touch Set Wallpaper, the wallpaper is changed and you are returned to the Home screen.

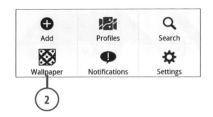

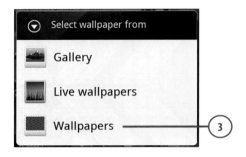

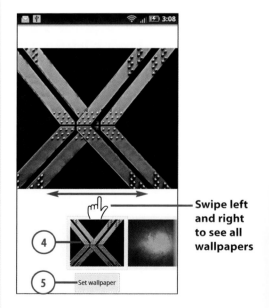

Swipe left and right to see all wallpapers

Wallpaper from Gallery Pictures

Instead of using specific wallpaper images, you can choose pictures from your Gallery to use as your DROID's wallpaper.

1. Touch or press the Menu button while on your Home screen.

2. Touch Wallpaper.

3. Touch Gallery.

4. Touch a picture in your gallery to use a your wallpaper.

5. Use the cropping box to adjust the way the picture is cropped before you use it as a wallpaper. Remember to expand the cropping box to the full picture size if you want to use the whole picture.

6. Touch to discard any changes and return to the Gallery to choose another picture.

7. Touch Save to use the picture as your wallpaper.

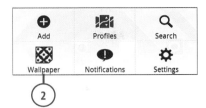

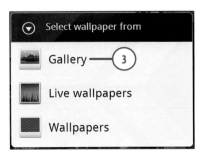

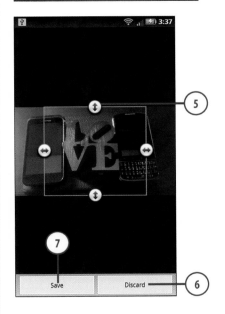

Live Wallpaper

Live wallpaper is wallpaper with
some intelligence behind it. It can be
a cool animation, or even an animation that keys off things such as the
music you are playing on your
DROID, or it can be something simple
such as the time. There are some very
cool live wallpapers, including one
that plays the game Pong to keep
the time.

1. Touch or press the Menu button
   while on your Home screen.

2. Touch Wallpaper.

3. Touch Live Wallpapers.

4. Touch a live wallpaper that you
   want to use.

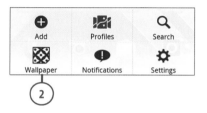

5. A preview of the live wallpaper is displayed.

6. Touch Settings to review and change the settings for the current live wallpaper. The settings you see on this screen are specific to the Live Wallpaper you have chosen.

7. Touch Set Wallpaper to use the live wallpaper.

Find More Wallpaper

You can find wallpaper or live wallpaper in the Android Market. Open Market and search for "wallpaper" or "live wallpaper." Read more on how to use the Android Market in Chapter 10, "Working with Android Applications."

One Touch to Set Wallpapers on the Incredible 2

On your DROID Incredible 2 (and other HTC Android phones), touch the icon to the right of the word Phone to set wallpaper, scenes, skins, and all other personalization settings.

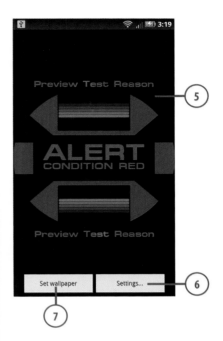

Touch to set wallpaper and other personalization settings on the Incredible 2

Using Scenes (HTC DROIDs Only)

If you have a DROID made by HTC, such as the HTC DROID Incredible 2, it comes preloaded with something called Scenes. Scenes are similar to themes on other smartphones and Windows computers. They change the wallpaper and add widgets to the different Home screens on your DROID. The idea is that you can switch scenes depending on your mood. You can use a work scene while at work and then switch it to a play scene after hours.

Changing the Scene

Depending on your mood, you might want to choose a new scene that changes the style and Home screen layout.

1. Touch the Personalize icon while on the Home screen.

2. Touch Scene.

3. Swipe left and right to see all Scenes.

4. Touch a Scene to select it.

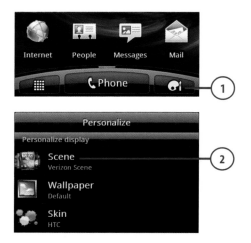

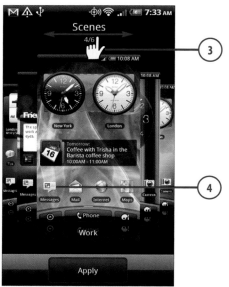

Creating Your Own Scene

After you have selected a scene you can modify the way it looks by moving icons and widgets around. You can also create new scenes that give you a blank canvas to place icons, widgets, and shortcuts just the way you like it.

1. Touch the Personalize icon while on the Home screen.

2. Touch Scene.

3. Touch the Menu button and then touch New.

4. Type the name of your new scene.

5. Touch Done.

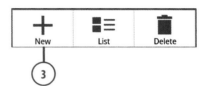

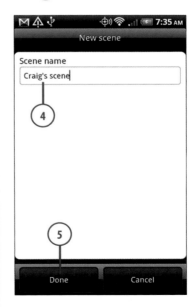

6. Your new scene is created and selected as the current scene automatically.

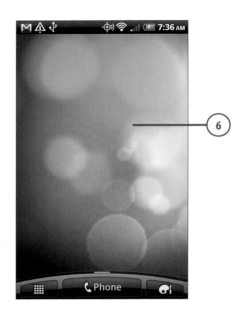

Using Profiles (Motorola DROIDs Only)

If you have the DROID 3, DROID Pro, or DROID X2, you can use profiles. Profiles are very similar to scenes on the DROID Incredible 2 that were discussed in the previous section. They enable you to quickly reset your DROID's Home screen layout.

Changing the Profile

Depending on your mood, you might want to choose a new Profile that changes the style and Home screen layout. Here is how.

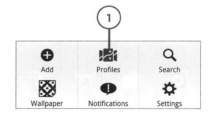

1. Press the Menu button and touch Profiles.

2. Touch a profile to select it.

Rename a Profile

If you don't like the names of the three existing profiles, touch and hold on one of the profiles. Then touch Rename in the pop-up menu.

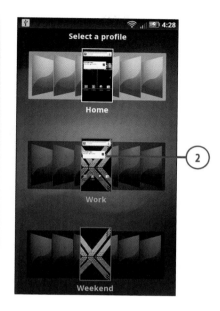

It's Not All Good

You Cannot Create Profiles

Unlike Scenes on HTC Android phones, you cannot create your own profiles. You can only use the three preinstalled profiles and rename them.

Using Skins (HTC DROIDs Only)

If you have a DROID made by HTC, such as the HTC DROID Incredible 2, it comes preloaded with something called skins. Skins enable you to change the look of the Sense bar at the bottom of the Home screen, the colors used for the status bar, and the color of the application title bars.

Changing the Skin

Depending on your mood, you might want to choose a different skin.

1. Touch the Personalize icon while on the Home screen.

2. Touch Skin.

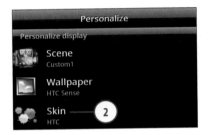

3. Swipe left and right to see all skins.

4. Touch a skin to use it.

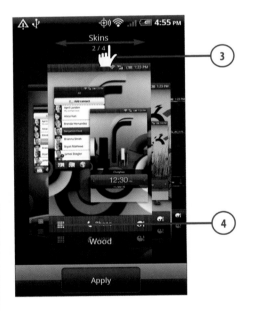

Changing Your Keyboard

If you find it hard to type on the standard DROID keyboard, or you just want to make it look better, you can install replacement keyboards. You can purchase or download replacement keyboards from the Android Market.

1. Launch Android Market to search for, purchase, or download a free keyboard replacement application. I chose one called Better Keyboard 8. Learn more about how to use the Android Market in Chapter 10.

2. Touch the Menu button and touch Settings.

3. Touch Language & Keyboard.

4. Touch the name of a keyboard (Better Keyboard 8, in this case) to choose your new keyboard.

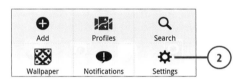

Do Your Research

When you choose a different keyboard in Step 4, the DROID gives you a warning telling you that nonstandard keyboards have the potential for capturing everything you type. Do your research on any keyboards before you download and install them.

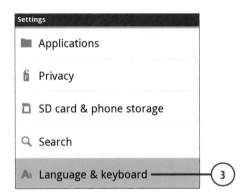

This input method may be able to collect all the text you type, including personal data like passwords and credit card numbers. It comes from the application Better Keyboard 8 (Gingerbread Edition). Use this input method?

OK   Cancel

Touch to accept the new input method

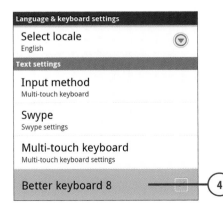

5. Touch Input Method.

6. Touch the name of your new keyboard to select it.

What Can You Do with Your New Keyboard

Keyboards you buy in the Android Market can do many things. They can change the key layout, change the color and style of the keys, offer different methods of text input, and even enable you to use an old T9 predictive input keyboard that you may have become used to when using an old "dumb phone" that only had a numeric keypad.

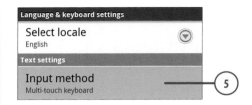

Better Keyboard 8 with a glowing skin

Swype® Keyboard

The Swype keyboard is a revolutionary keyboard that enables you to swipe your finger over the keyboard to type as opposed to touching each key individually. Theoretically, this enables you to type more quickly. It's cool whether it speeds your typing or not. The Swype keyboard is preinstalled on the DROID 3 and DROID X2. Unfortunately, it cannot be downloaded from the Android Market, which means it's only available on the DROID 3 and DROID X2. Here is how to switch to the Swype keyboard on the DROID 3 and DROID X2.

1. Press the Menu button and touch Settings.

2. Touch Language & Keyboard.

3. Touch Input Method.

4. Touch Swype.

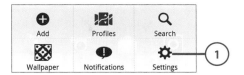

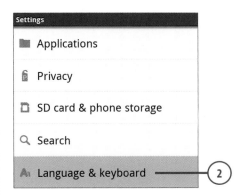

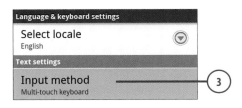

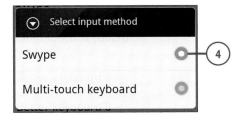

BASICS OF SWYPE

There is a full tutorial on how to use Swype on the DROID X, but to get you going, here are the basics.

- Type a word by swiping your finger across all the keys that make up the word and then lifting your finger.

- To type a double letter, circle the letter.

- To type an uppercase letter, swipe off the keyboard after swiping over the letter, then swipe over the remaining letters in the word.

- Sometimes you type a word that is too similar to another word. At that point, the Swype keyboard displays a choice of possibilities. Touch the word you want.

Adding Widgets to Your Home Screens

Some applications that you install come with widgets that you can place on your Home screens. These widgets normally display real-time information such as stocks, weather, time, and Facebook feeds. Your DROID also comes preinstalled with some widgets. Here is how to add and manage widgets.

Adding a Widget

Your DROID should come preinstalled with some widgets, but you might also have some extra ones that have been added when you installed other applications. Here is how to add those widgets to your Home screens.

1. Touch the Menu button and touch Add or the plus symbol.

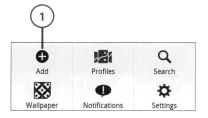

Adding Widgets on Your DROID Incredible 2

If you have a DROID Incredible 2, to add widgets, touch the Personalize icon on the bottom right of the Home screen and then touch Widgets. The rest of the steps are the same.

---

2. Touch either Motorola Widgets or Android Widgets.

Moto or HTC Widgets

Because Android is so customiz-able, this screen might look differ-ent. For example, if you have a DROID X2, the screen shows Motorola Widgets and Android Widgets. If you have a DROID Incredible 2, this screen shows HTC Widgets and Android Widgets.

3. Touch the widget you want to add.

HTC Widgets

If you have an HTC DROID, on the Widgets screen you see a Get More HTC Widgets option. Touch this to see a list of HTC-specific widgets that you can download and use on your DROID.

4. The widget is added to the Home screen.

Moving a Widget

Sometimes you want to reposition your widgets or move them to other screens. Here is how.

1. Touch and hold the widget you want to move.

2. Move your finger around the screen. As you move, a green box follows to indicate where you can move the widget to.

3. You can also drag the widget off the screen to the left or right to place it on an adjacent screen.

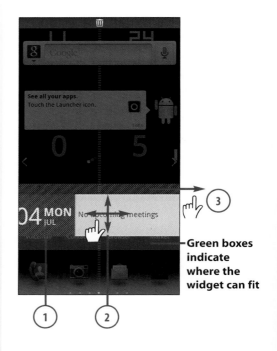

Green boxes indicate where the widget can fit

Removing a Widget

When you become tired of a certain widget or need to make space for a new widget you can remove widgets from your Home screen.

1. Touch and hold the widget you want to remove.

2. Drag the widget to the word Remove. Depending on the model of DROID, you might see a trash icon instead of the word Remove, and it might be on the top or bottom of the screen.

3. The widget is removed from the screen, but it's still on your DROID for future use—just use the steps from the "Adding a Widget" task.

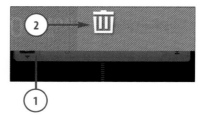

Region and Language

If you move to another country or want to change the language used by your DROID, you can do so with a few touches.

1. From the Home screen, touch the Menu button and touch Settings.

2. Touch Language & Keyboard.

3. Touch Select Locale.

4. Touch the language you want to switch to. The number of languages listed here depend on the carrier and DROID model.

5. Your DROID instantly switches to using this new language.

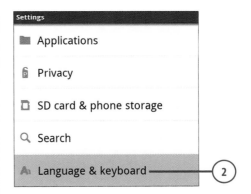

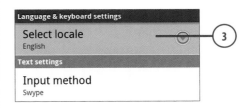

Who Obeys the Language?

When you switch your DROID to use a different language you immediately notice that all standard applications and the DROID menus switch to the new language. Even some third-party applications honor the language switch. However many third-party applications ignore the language setting on the DROID. So you might open a third-party application and find that all of its menus are still in English.

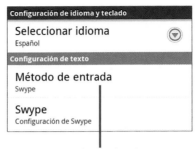

Screen information is now displayed in the new language

Accessibility Settings

Some DROIDs include built-in settings to assist people who might otherwise have difficulty using some features of the device. The DROID has the ability to provide alternative feedback such as vibration, sound, and even speaking of menus.

1. While on the Home screen, touch the Menu button and then touch Settings.

2. Scroll down and touch Accessibility.

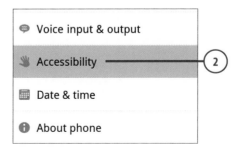

DROID Incredible 2 Has No Accessibility Apps

If you have a DROID Incredible 2, it does not come preloaded with any accessibility apps. However, when you touch Accessibility in Step 2, you are prompted to download a screen reader from the Android Market. While there, you can download many other free accessibility apps. After you have downloaded them, they appear in the Accessibility screen.

3. Touch Accessibility to enable the accessibility services. The green check mark indicates the services are turned on.

4. Select which accessibility apps you want to enable.

5. Touch to enable a feature where the power button can be used to end phone calls.

What Do the Accessibility Apps Do?

Zoom Mode provides a movable box on the screen that zooms in on whatever you place it over.

SoundBack plays sounds as you navigate your DROID. SoundBack adds audible cues as you navigate menus on your DROID. For example if you touch the search area, you hear a ping sound.

KickBack makes your DROID vibrate as you navigate menus and press buttons.

TalkBack speaks menus, labels, and names as you navigate on your DROID.

No accessibility related applications found

You do not have any accessibility related applications installed.

You can download a screen reader for your device from Android Market.

Click OK to install the screen reader.

OK   Cancel

Touch to download accessibility apps

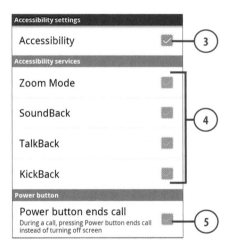

Control Volume of SoundBack and TalkBack

You can control the volume of SoundBack and TalkBack because they fall under the category of Media playback. However it is impossible to control the volume with just SoundBack enabled because the audible cues are too quick. A good trick is to enable TalkBack and, while it is reading something, adjust the volume by using the physical volume keys on the side of your DROID. After you have the desired volume, you can disable TalkBack.

Search Settings

When you search from the Home screen on your DROID, you are searching the Internet as well as content on your phone such as your contacts and browser bookmarks. You can control what to search, by specifying the searchable items.

1. While on the Home screen, touch the Menu button and then touch Settings.

2. Scroll down and touch Search.

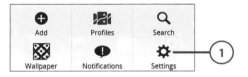

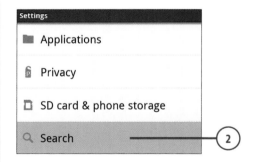

3. Touch to clear your previous search terms.

4. Touch Searchable Items to select what to search on the Web and which apps on your DROID you want to include in searches.

5. Touch to select whether you want your location sent when you search the Web, whether to enable suggestions when you search, and whether to show your personalized search history.

In this chapter, you learn how to maintain your DROID and solve problems. Topics include the following:

→ Updating Android
→ Battery Optimization
→ Find battery hungry applications
→ Caring for your DROID

12

Maintaining DROID and Solving Problems

Every so often Google releases new versions of Android that have bug fixes and new features. In this chapter you find out how to upgrade your DROID to a new version of Android and how to tackle common problem-solving issues and general maintenance of your DROID.

Updating Android

New releases of Android are always exciting because they add new features, fix bugs, and tweak the user interface. Here is how to update your DROID.

Update Information

Updates to Android are not on a set schedule. Complicating matters is the fact that there are different phones running Android so your friend who has a different Android phone might get an update before you see the update message on your DROID. The update messages appear as you turn on your DROID, and they remain in the notification bar until you install the update. If you touch Install Later, your DROID reminds you that there's an update every 30 minutes. Sometimes people like to wait to see if there are any bugs that need to be worked out before they update, so it is up to you.

1. If you receive a notification that there is a system update, touch Install Now.

Manually Check for Updates

If you think there should be an update for your DROID, but have not yet received the onscreen notification, you can check manually by touching Menu, Settings, About Phone, and System Updates. If there are updates, they are listed on this screen.

2. Touch Restart & Install. Your DROID updates and reboots.

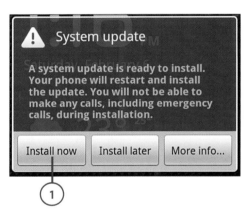

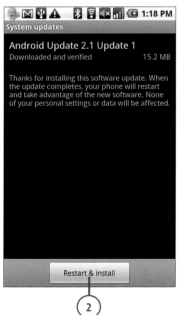

Battery Optimizing

The battery in your DROID is a lithium ion battery that provides good battery life when you take care of it. Changing the way you use your DROID helps prolong the battery's life, which gives you more hours in a day to use your phone.

Looking After the Battery

There are specific steps you can take to correctly care for the battery in your DROID. Caring for your battery enables it to last longer.

1. Try to avoid discharging the battery completely. Fully discharging the battery too frequently harms the battery. Instead, try to keep it partially charged at all times (except as described in the next step).

2. To avoid a false battery level indication on your DROID, let the battery fully discharge about every 30 charges. Lithium-ion batteries do not have "memory"-like older battery technologies; however, the battery meter is the problem.

3. Do not leave your DROID in a hot car or out in the sun anywhere, including on the beach, as this can damage the battery and make it lose charge quickly.

4. Consider having multiple chargers. For example, you could have one at home, one at work, and maybe one at a client's site. This enables you to always keep your DROID charged.

Automatically Ending Applications—DROID 3, DROID X2, Pro, and CHARGE

Your DROID 3, X2, Pro, and CHARGE have an application called Task Manager pre-installed. The Task Manager enables you to see which applications must automatically close after 2 minutes after your DROID's screen goes blank. This helps to save on battery life.

Where Do I Find Task Manager on My DROID Incredible 2?

Your DROID Incredible 2 does not come with the Task Manager app pre-installed, however if you use the Android Market application, you can find hundreds of these apps when you search for "task killer" or "task manager." You don't need to pay money for one because there are many free task killers or task managers. See Chapter 10, "Working with Android Applications," for more information on how to use the Android Market.

1. Touch the Task Manager icon.

2. Touch the check box next to any application you want to close or automatically end.

3. Touch End Apps to end the selected applications now.

4. Touch Add to Auto-end to add the selected applications to the Auto-end list. Your DROID automatically closes apps on the auto-end list two minutes after your DROID's screen goes blank. Even if you restart your DROID, any apps you have added to the Auto-End list remain there.

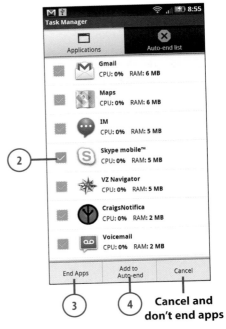

Cancel and don't end apps

Quicker Way to End Apps

If you only want to end one app or only want to add one app to the auto-end list, instead of touching the check box just touch the name of the app to see a pop-up window. Touch End Application or Add to Auto-end

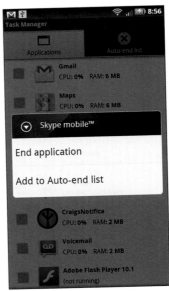

List.

Get Warned About High CPU or Memory Usage

You can also get the Task Manager to warn you if your DROID is experiencing high CPU or memory usage. Many times this means that you have a "runaway app" that needs to be ended. To do this, with no applications selected, press the Menu button and touch Settings. Check the box labeled Notification.

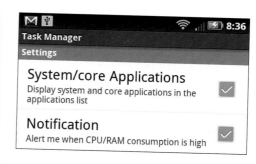

Battery Optimization— DROID 3, X2, Pro, CHARGE

You can further control how your battery is used by applications and processes, and even utilize battery modes to squeeze more life out of a battery.

1. Touch or press the Menu button

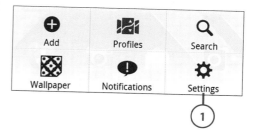

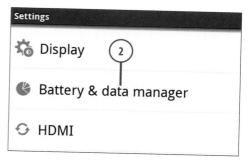

while on the Home screen and touch Settings.

2. Touch Battery & Data Manager.

3. Touch Battery Usage to see which applications are using the most battery.

4. Touch to select the battery mode. This enables you to select between different battery modes such as Nighttime Saver, Performance Mode, and Maximum Battery Saver. These modes control how the battery is used.

5. Touch a battery mode to select it.

6. Touch to select your custom battery mode.

7. Touch to edit your custom battery mode.

Why Not Just Change These Settings Myself?

You could turn your radios on or off, control when applications synchronize data, or control the screen brightness and timeout yourself, but the battery modes

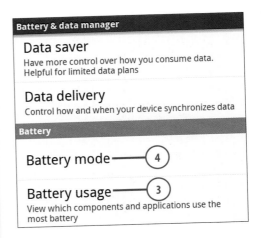

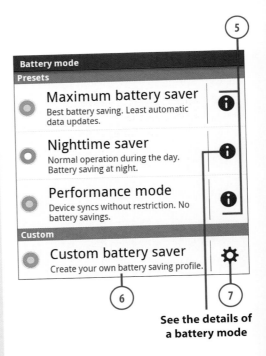

See the details of
a battery mode

have been designed for optimum battery performance so it is better to leave them.

Battery Optimization— DROID Incredible 2

You can further control how your battery is used by applications and processes, and even utilize battery modes to squeeze more life out of a battery.

1. Touch or press the Menu button while on the Home screen and touch Settings.

2. Touch Power.

3. Touch to enable the Power Saver feature.

4. Touch to select at what battery power level the Power Saver feature must activate.

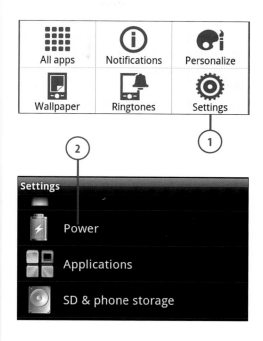

5. Touch to be notified when the Power Saver feature activates.

6. Touch to adjust what the Power Saver feature does when it is activated.

What Is the Point of the Power Saver Feature?

The point of the Power Saver feature is to limit what your DROID Incredible is doing when the battery is low on power. It essentially squeezes extra time out of the battery by turning off unused radios, shortening the screen timeout and brightness, disabling background data synchronization, disabling

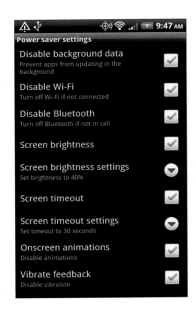

screen animations, and turning off any vibrate feedback. You can however, control which features are disabled when power savings kicks in. For example, you can set your DROID not to disable Bluetooth.

Dealing with Misbehaving

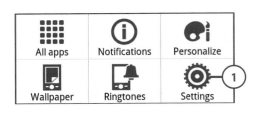

Applications

Sometimes applications misbehave and slow down your DROID. You can use a task-killer application as we discussed in the "Battery Optimization" section, or you can use the DROID's built-in feature to forcefully close an application.

1. Touch the Menu button and touch Settings.

2. Touch Applications.

3. Touch Manage Applications.

4. Touch Running.

5. Touch the misbehaving application.

6. Touch Force Stop. The screen does not update, but the application stops running. If you go back to the list of applications you see that the application is no longer visible.

When to Force Stop an Application

After you have been using your

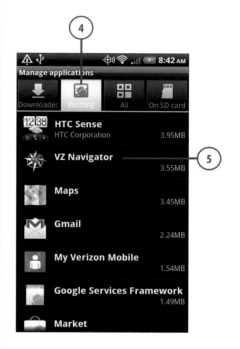

DROID for a while, you'll become familiar with how long it takes to do certain tasks such as typing, navigating menus, and so on. If you notice your DROID becoming slow or not behaving the way you think it should, the culprit could be a new application you recently installed. Because Android never quits applications on its own, that new application continues running in the background and causing your DROID to slow down. This is when it is useful to use force stop.

Caring for DROID's Exterior

Because you need to touch your DROID's screen to use it, it picks up oils and other residue from your hands. You also might get dirt on other parts of the phone. Here is how to clean your DROID.

1. Wipe the screen with a microfiber cloth. You can purchase these in most electronic stores, or you can use the one that came with your sunglasses.

2. To clean dirt off other parts of your phone, wipe it with a damp cloth. Never use soap or chemicals on your DROID as they can damage it.

3. When inserting the Micro-USB connector, try not to force it in the wrong way. If you damage the pins inside your DROID, you will not be able to charge it unless you have the dock.

Getting Help with Your DROID

There are many resources on the Internet where you can get help with your DROID.

1. Visit the Official Google website at http://www.android.com.

2. Check out some Android blogs:

   • Android Central at http://www.androidcentral.com/

   • Android Guys at http://www.androidguys.com/

   • Androinica at http://androinica.com/

3. Contact me. I don't mind answering your questions, so visit my official *My DROID* site at http://www.CraigsBooks.info.

Index

FREE Online Edition

Your purchase of *My DROID™ Second Edition* includes access to a free online edition for 45 days through the Safari Books Online subscription service. Nearly every Que book is available online through Safari Books Online, along with over 5,000 other technical books and videos from publishers such as Addison-Wesley Professional, Cisco Press, Exam Cram, IBM Press, O'Reilly, Prentice Hall, and Sams.

SAFARI BOOKS ONLINE allows you to search for a specific answer, cut and paste code, download chapters, and stay current with emerging technologies.

Activate your FREE Online Edition at
www.informit.com/safarifree

> **STEP 1:** Enter the coupon code: VCCQYYG.

> **STEP 2:** New Safari users, complete the brief registration form. Safari subscribers, just login.

If you have difficulty registering on Safari or accessing the online edition, please e-mail customer-service@safaribooksonline.com